WEATHER
of
ONTARIO

Phil Chadwick
Bill Hume

Lone Pine Publishing

© 2009 by Lone Pine Publishing
First printed in 2009   10 9 8 7 6 5 4 3 2 1
Printed in China

All rights reserved. No part of this work covered by the copyrights hereon may be reproduced or used in any form or by any means—graphic, electronic or mechanical—without the prior written permission of the publisher, except for reviewers, who may quote brief passages. Any request for photocopying, recording, taping or storage on information retrieval systems of any part of this work shall be directed in writing to the publisher.

The Publisher: Lone Pine Publishing
10145 – 81 Avenue
Edmonton, AB, Canada   T6E 1W9

Website: www.lonepinepublishing.com

Library and Archives Canada Cataloguing in Publication

Chadwick, Phil, 1953-
    Ontario weather / Phil Chadwick, Bill Hume.

Includes index.
ISBN 978-155105-608-1

    1. Ontario--Climate. I. Hume, Bill, 1949- II. Title.

QC985.5.O6C43 2009   551.69713   C2009-900315-5

Editorial Director: Nancy Foulds
Project Editor: Gary Whyte
Editorial: Wendy Pirk, Sheila Quinlan, Gary Whyte
Photo Coordination: Gary Whyte, Randy Kennedy
Production Manager: Gene Longson
Book Layout & Production: Rob Tao
Maps & Charts: Volker Bodegom, Rob Tao
Illustrations: Michael Cooke, Gerry Dotto, Megan Fischer, Trina Koscielnuk, Rob Tao
Cover Design: Gerry Dotto

Front cover photograph by: JupiterImages Corporation
Title Page: "Jet Stream Cirrus" by Phil Chadwick

Photo Credits: Every effort has been made to accurately credit copyright owners. Any errors or omissions should be directed to the publisher. The photographs in this book are reproduced with the kind permission of the copyright owners. All photo credits are located on p. 239.

We acknowledge the financial support of the Government of Canada through the Book Publishing Industry Development Program (BPIDP) for our publishing activities.

PC: 16

Table of Contents

Introduction 5

Chapter 1   The Atmosphere 10
The Composition of the Atmosphere, 11 • Air Masses and Fronts, 20

Chapter 2   Clouds 25
Cloud Classification, 30 • Clouds with Vertical Development, 47

Chapter 3   Precipitation 54
The Raindrop Formation Process, 56 • Types of Hydrometeors, 57 • Virga, 64 • Precipitation Transitions, 65

Chapter 4   Atmospheric Electricity 66
Lightning, 66 • The Cloud Electrification Process, 67 • Generation of a Lightning Stroke, 70 • Areas of Lightning in Ontario, 76 • Effects of Lightning Flashes, 77 • Lightning Protection, 78 • Passive Electrical Phenomena, 79

Chapter 5   Winds 83
Lake Breezes, 84 • Land Breezes, 85 • Terrain Effects, 86 • Jets, 87 • Convection-Induced Winds, 90 • Tornadoes, 92 • Onshore Winds and Shoreline Convergence, 99

Chapter 6   Optics and Acoustics 100
Blue Sky, 100 • Rainbows, 102 • Ice Crystal Phenomena, 105 • Mirages, 107 • Looming, 108 • Bright Water Horizons, 109 • Crepuscular Rays, 109 • Twinkling Stars, 109 • Other Optical Effects, 109 • Acoustical Effects, 110

Chapter 7   Ontario's Climate 112
Controls and Influences, 112 • Temperature, 123 • Combinations of Weather Elements, 132 • Precipitation, 134 • Drought, 138 • Sunshine, 141

Chapter 8   Storms 142
Winter Storms, 144 • Summer Storms, 152 • Autumn Storms, 160

Chapter 9   Observing the Weather 161
Observation Systems, 162 • Communication of Meteorological Data, 180 • Data Quality Control, 180

Chapter 10   Forecasting the Weather 181
Numerical Weather Prediction, 182 • The Forecast Products, 186 • The Role of the Forecaster, 190

Chapter 11   Air Quality 193
Emissions to the Atmosphere, 194 • Pollution Movement in the Planetary Boundary Layer, 198 • Chemical Transformation in the Atmosphere, 202 • Air Quality as a Societal Issue in Ontario, 205

Chapter 12   The Changing Atmosphere 207
Stratospheric Ozone, 208 • Climate Change, 211 • Future Climate, 221 • A Plan for All Seasons, 227

Glossary 228
Resources 232
Index 234

Dedication

To Linda and our family, Janice and Keith, who have made this wonderful life a real adventure—and the best is yet to come. And the quiet environmentalists like my father, Nelson, and good friend, Paul. The efforts of like-minded individuals give hope to the future.

Acknowledgements

Many thanks to all the people who helped bring this book together. I would especially like to thank my co-author Bill Hume whose *Weather of Alberta* provided an excellent background text and model for this book. The staff of Environment Canada and several Ontario Provincial Agencies provided key scientific data, research and suggestions. These countless friends include William (Bill) Laidlaw, Bill Burrows, David Sills, Dave Broadhurst, Geoff Coulson, Wade Szilagyi, Duncan Fraser, Norman Donaldson, Joan Klaassen, Sharon Fernandez, Bill Richards and Amir Shabbar to name just a few.

A very special thank you to David Sills for taking the time to review the complete manuscript and for providing numerous helpful and useful comments—the book is definitely better for it!

Thanks also to my weather-spotter CANWARN friends who contributed images from their personal libraries. George Kourounis and Dave Patrick came through with images I lacked. These volunteers are a vital part of the safety net that helps to maintain awareness about severe weather and to safeguard the public.

The direction and guidance shown by Nancy Foulds, Wendy Pirk, Gary Whyte, Ken Davis and the rest of the staff at Lone Pine Publishing was much appreciated.

I would especially like to thank my meteorologist friends Alister Ling and Bob Jones who suggested that I take on this project—as if I didn't have enough to do.

Finally, I would like to express my sincere gratitude for the opportunity to spend an entire career investigating the many mysteries of the weather while serving the public and looking out for their safety and security. There is weather every day and every day was a fun challenge to get it right—or at least close! Thank you …

—*Phil Chadwick*

"Smoke Lake Cumulus" by Phil Chadwick

Introduction

Weather is important! Everyone likes to talk about the weather and everyone is interested in it. Even though most of us now lead urban lives, weather is still the driving force that influences our daily activities. Daily weather, added up and averaged, becomes climate. Weather is what we *get* whereas climate is what we *expect*. And both are changing.

The weather in each province is certainly unique and—as local meteorologists will claim—the most difficult to forecast. Probably everyone is correct. Weather changes like a chameleon and successful prediction is very challenging. The varying weather and seasons provide a beautiful backdrop to the special environment that Canada and Ontario enjoy. Weather can change in the span of minutes or simply down the road. If you don't like the weather, wait ten minutes or visit the next county.

Weather makes the news and has become part of Ontario folklore and the social fabric. The Ice Storm of '98, Hurricane Hazel of '54 and the Barrie Tornado of '85 all figure prominently. The Black Sunday Storm of November 1913 sank 34 ships on Lakes Erie and Ontario and was one of the top weather events from the 20th century. Weather inspired Tom Thomson and the Group of Seven. Musicians can't stop composing about the weather. Future weather events will continue to write weather history. And those events could even happen today!

Introduction

Ecoregions of Canada

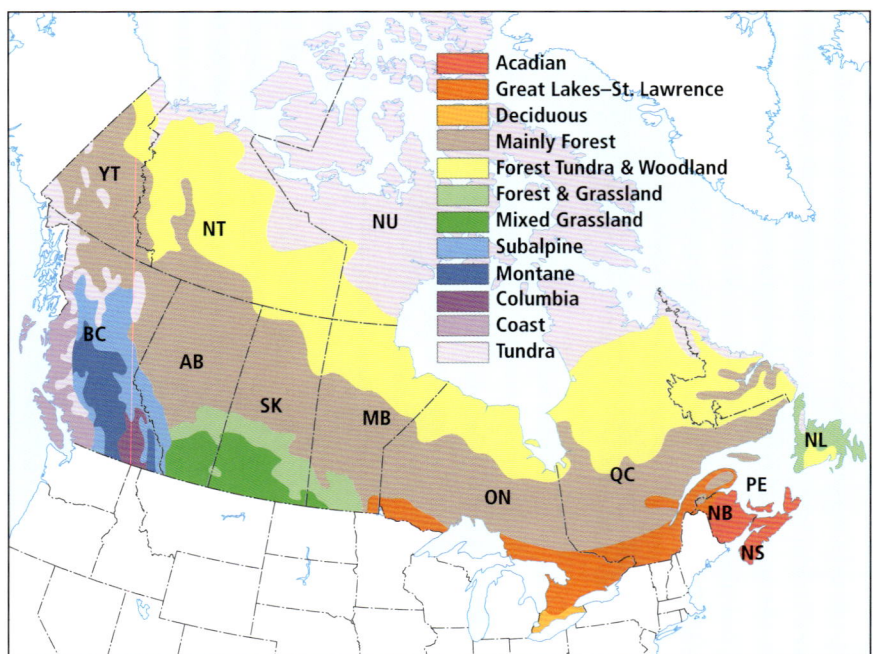

Other provinces might chuckle at the Toronto snowstorm that required the troops to get things moving again. However, sometimes it is not just the weather but the fragile and complex infrastructure it affects. Modern society has certainly evolved but instead of being more resilient to the effects of the environment, society has actually become more vulnerable. As a result, weather and climate prediction have become more essential.

Ontario, the most populated province in Canada, gets its share of weather. Air masses from all directions battle it out over Ontario with the Great Lakes adding some seasoning to the weather that results. Some weather is even manufactured on site. The Great Lakes, the Niagara Escarpment, the James Bay Lowland and the Algonquin Highlands all have huge impacts on the air masses as they arrive.

The moderating effect of the Great Lakes is certainly the largest influence on Ontario weather. These inland seas tend to delay both the warming of the land in summer and the cooling of the land in winter. They provide heat and moisture to the air in winter, creating severe snowsqualls which can bring metres of fluffy flakes to the onshore areas and ski belts while dropping the visibility to less than the hood of your car. The same lakes can create cool breezes that focus summer thunderstorms. The shape and orientation of the shorelines dictate exactly where all of this weather occurs.

Introduction

Ecological Regions of Ontario

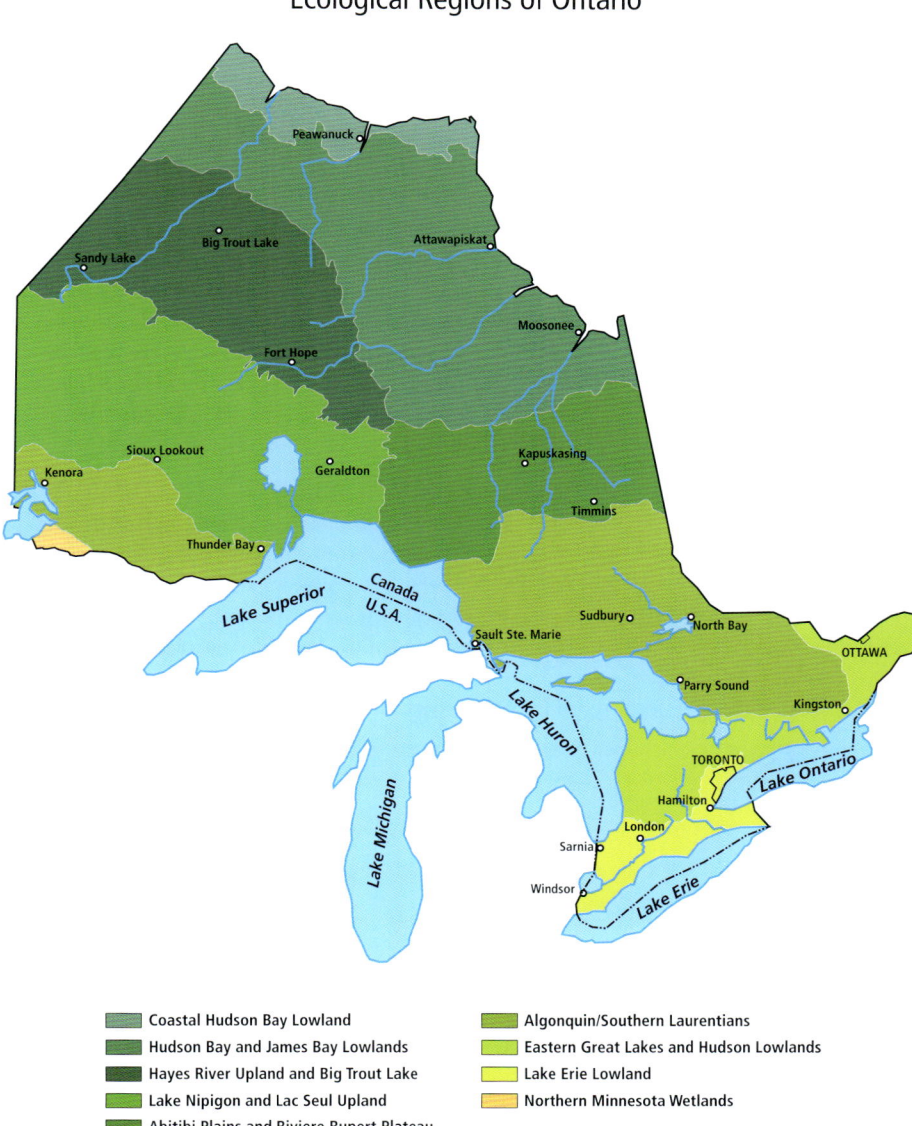

- Coastal Hudson Bay Lowland
- Hudson Bay and James Bay Lowlands
- Hayes River Upland and Big Trout Lake
- Lake Nipigon and Lac Seul Upland
- Abitibi Plains and Riviere Rupert Plateau
- Algonquin/Southern Laurentians
- Eastern Great Lakes and Hudson Lowlands
- Lake Erie Lowland
- Northern Minnesota Wetlands

Introduction

Low-pressure areas typically intensify as they cross the Great Lakes in fall and winter. Heat, moisture and a lack of friction from the inland seas combine to wind up severe storms that invoke awe and fear in mariners. The *Edmund Fitzgerald*, immortalized by the haunting chords of Gordon Lightfoot is just one of more than 10,000 shipwrecks that lie on the bottom of the Great Lakes.

The tundra of the James Bay Lowland is no barrier to the frigid air of the Arctic pipeline and northwesterly winter winds. In contrast, southerly summer downslope winds can make the shores of James Bay the hottest in the province.

These influences all combine to ensure that Ontario does not win many of the weather extreme awards. Southern Ontario does excel in hot, humid summer weather with poor air quality and thunderstorms. Portions of Northern Ontario do well in the cold winter categories. While Alberta might win the prize as the province with the "most comfortable weather" and Nunavut wins in the toughest weather category, Ontario is the winner of the prestigious "all seasons" award. This award means that Ontario gets a little bit of almost every type of Canadian weather but generally nothing too extreme. Ontario weather is very interesting without being overly dangerous. To quasi-quote Goldilocks " 'Ahhh, this *weather* is just right,' she said happily, and she *soaked* it all up."

The Science of Meteorology

Meteorology is the science that defines how the atmosphere behaves. This science is based on physics and mathematics, and it quickly becomes complex, if not downright confusing. This book explains the

Province-Territory for All Seasons

Rank	Province/Territory	Points
1	Ontario	20
2	Manitoba	18
3	Saskatchewan	12
4	British Columbia	11
5	Quebec	9
6	Newfoundland and Labrador	7
7	New Brunswick	4
8	Prince Edward Island	4
9	Alberta	4
10	Nova Scotia	2
11	Yukon Territory	2
12	Northwest Territories	2
13	Nunavut	1

Introduction

behaviour of the atmosphere with narrative descriptions and diagrams, but does not include a single equation. It is said that a picture is worth a thousand words, so the images included in these pages should sufficiently illustrate meteorology's many scientific complexities. Those interested in a more rigorous treatment of the science are better off referring to any number of scientific texts, some of which are included in a list of references at the end of this book. There are also many excellent web sites that can be investigated. The Environment Canada site (www.msc.ec.gc.ca/weather) provides much information on atmospheric science. Weather forecasts are available at the Canadian Weather Network site (www.theweathernetwork.com) and at the Environment Canada site (www.weatheroffice.gc.ca).

The science of meteorology has made great strides forward in the prediction of the atmosphere. These predictions are especially skillful in the mid and upper levels of the atmosphere. Interpreting these predictions to weather where people live and at the time and space scales of interest is still a large challenge.

This book will provide you with an understanding of weather and many of the associated phenomena, with an emphasis on Ontario and its varied climate. Within these pages, we discuss the atmosphere—its structure and how it works—as well as many atmospheric phenomena.

Southern Ontario gets a lot of thunderstorms so there is an important section on convective storms and their characteristics. And no discussion of storms would be complete without a mention of atmospheric electricity.

Because weather has such a huge impact on our day-to-day lives, people throughout time have been keeping an eye on the skies. Weather observation is a science that has a long history: records of formal weather observation in Canada date back to the 18th century. The earliest regular measurements in Ontario began on January 1, 1840, on the lands now known as Old Fort York in Toronto. In this book, descriptions of the measurement systems and equipment used to observe weather in the province are provided.

Venturing beyond the day-to-day weather phenomena, the final chapters examine two broad atmospheric issues: air quality and how the atmosphere deals with air pollution; and the long-term future of our atmosphere, including changes that we have induced on the protective ozone layer and the important issue of climate change.

Weather may be complex, but it's *not* chaotic. The weather actually plays by strict meteorological rules. The goal is to learn these rules by reading the clouds and weather patterns. By reading them, one can understand the atmosphere—understand weather. Weather in all its complexity is beautiful and is with us every day. Enjoy the challenge of understanding the weather and marvel at the magnificence that surrounds us.

I sincerely hope you look up more often at the weather of Ontario, and that you enjoy this book!

Chapter 1: The Atmosphere

> Some are weather-wise,
> some are otherwise.
> —Benjamin Franklin

Compared to the dimension of the earth, the atmosphere is just a thin skin of gas. But it is complex; it contains a number of constituents, has a complicated structure and has many protective properties without which life as we know it could not survive. As we note in pictures taken from space, the blue atmosphere has considerable associated structure, which is visible because of the presence of clouds. When observed with time-lapse photography, movement and changes in the cloud patterns suggest that there is indeed some structure. There may be large areas with organized shape. White cloud masses swirl, expand and shrink, while cloud bands in the shape of lines and arcs move across the earth's surface.

Although the motion may appear chaotic, it actually closely follows paths and patterns that are described by rules governing the complex science of fluid dynamics. These are the same laws that describe the motion of fluids flowing in channels and pipes, or the motion of air around an aircraft wing. These motions all have a chaotic nature to them—they have some degree of associated turbulence. But the flow mostly behaves in an explainable manner, and its motion can be predicted.

The Atmosphere

The Composition of the Atmosphere

The composition of the atmosphere is 21 percent oxygen and 78 percent nitrogen. The remaining one percent includes a wide variety of other gases, including water vapour, carbon dioxide, the inert gas argon and a host of others, some naturally occurring and some human made. Oxygen and nitrogen are well mixed throughout the atmosphere. Many of the other gases are less uniformly distributed, a result of their tendency to chemically react with other gases, to change phase, or to remain in the region of the atmosphere near their emission source. Gravity keeps the gaseous envelope to within a depth of a few hundred kilometres of the earth's surface. The interaction between molecules pushes them apart, preventing the gas from collapsing into a lump of oxygen and nitrogen at the surface. This interaction is known as internal pressure. The molecules constantly collide with each other and with other objects. Near sea level, at temperatures near 20° C, the individual molecules move at about 500 metres per second. Molecules in the lower atmosphere are dense—near the earth's surface, 1 cubic centimetre contains about 25 sextillion molecules (or 25 followed by 21 zeroes!).

To picture an analogy for pressure, imagine yourself throwing balls at a bathroom scale. Each time a ball hits the scale, the scale registers a slight reading, which depends on the speed and weight of the ball that was thrown. When many balls are thrown over a fixed period of time, say one second, the weight indicator on the scale does not have time to return to zero—a steady reading, or pressure of the colliding balls, is displayed. It is the same with the atmosphere. Over a 1-square-centimetre surface, there are hundreds of millions

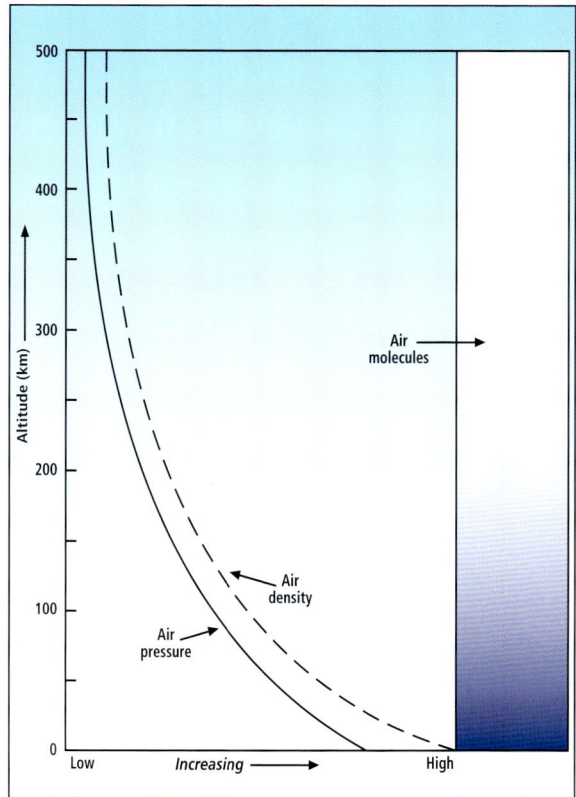

Variation of Atmospheric Pressure and Density with Height

Fig. 1-1

The Atmosphere

of molecules, each with a tiny mass, striking per second. This molecular pressure pushing in all directions, including upward, counteracts the downward pull of gravity. The mass of all the molecules in a vertical column with a surface area of 2.5 by 2.5 centimetres extending from sea level to the top of the atmosphere is about 6.7 kilograms. This internal pressure, measured as a force per unit area, pushes back the earth's gravitational attraction to the molecules. In atmospheric science, the pressure force is called the millibar (mb).

The higher one goes in the atmosphere, the fewer molecules there are, and the weight of the molecules, as well as the corresponding internal pressure, is reduced. There is a mathematical equation that defines how atmospheric pressure decreases with height. This equation shows (and measurement confirms) that at a height of 5 kilometres, atmospheric pressure is reduced to about 50 percent of its value at sea level, and at a height of 10 kilometres, to about 30 percent.

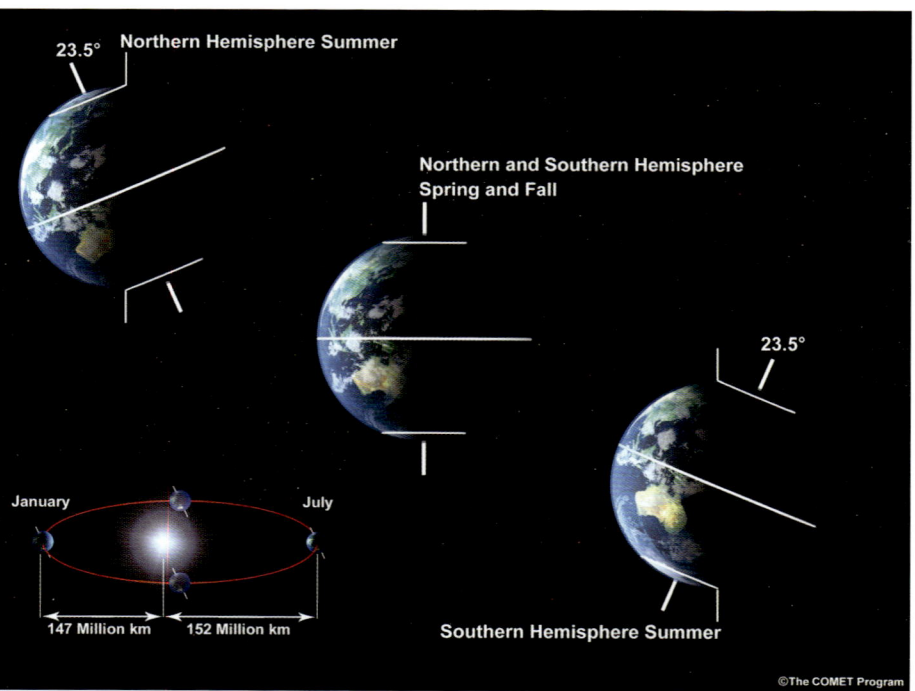

Fig. I-2 The tilt of the earth's axis of rotation gives us seasonal changes.

The Atmosphere

Solar Energy Reflected by Earth's Surfaces

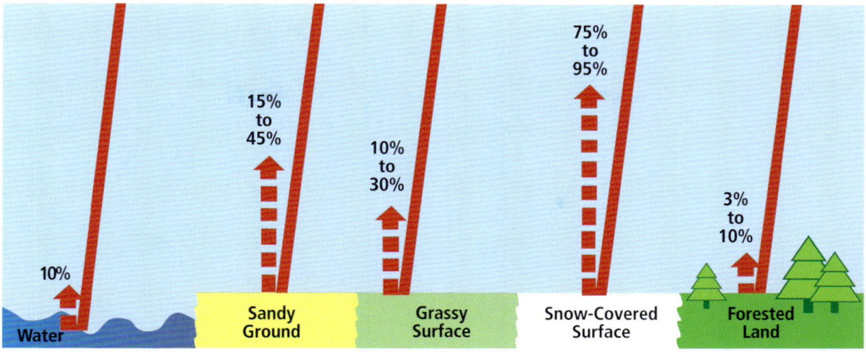

Fig. 1-3 Different surfaces reflect varying amounts of the sun's radiation.

Solar Input and its Variation

The sun is the primary driver of weather on earth, but it is the tilt of the earth's axis of rotation that gives us the seasonal changes. This angle of inclination (23 degrees) remains constant as the earth proceeds on its annual trek around the sun. The orbit is not quite circular—it is elliptical. When the earth's axis is tilted away from the sun, and Ontario is in the middle of winter, the earth is at its closest point to the sun. Regions north of 66.5° N latitude receive no winter sunlight. The most northern part of Ontario, at Fort Severn, is about 56° N, so no part of Ontario is plunged into total darkness in winter or total daylight in summer when the axis is tilted toward the sun.

The sun's radiation covers the full electromagnetic spectrum, from the very short x-rays, through the visible light rays to the much longer radio wave frequencies. However, the amount of radiation differs across the spectrum. About 40 percent of the sun's energy is emitted in the visible and near infrared frequencies. The amount of solar radiation that reaches the earth's surface is less than what is incident at the top of the atmosphere. Ozone molecules in the high atmosphere absorb most of the ultraviolet energy, whereas much of the incoming infrared energy is absorbed in the lower levels of the atmosphere.

Clouds and snow surfaces can reflect up to 95 percent of incident radiation, whereas water surfaces reflect only about 10 percent. Sand, grass surfaces and forests reflect between 10 and 30 percent. These reflections depend on the angle of incidence of the radiation—the smaller the angle of incidence with the surface, the greater the amount of reflection. Overall, 43 percent of the solar radiation entering the top of the atmosphere is reflected and scattered back to space. Some absorption by dust and gases, including water vapour, further depletes the incoming radiation, with the result that about 40 percent is absorbed by the earth's surface. All bodies that absorb energy are heated, and these warm bodies in turn emit radiation at the infrared frequency. Some of this energy makes it back out into space if it is not intercepted

The Atmosphere

Energy Transfers in the Earth-Atmosphere System

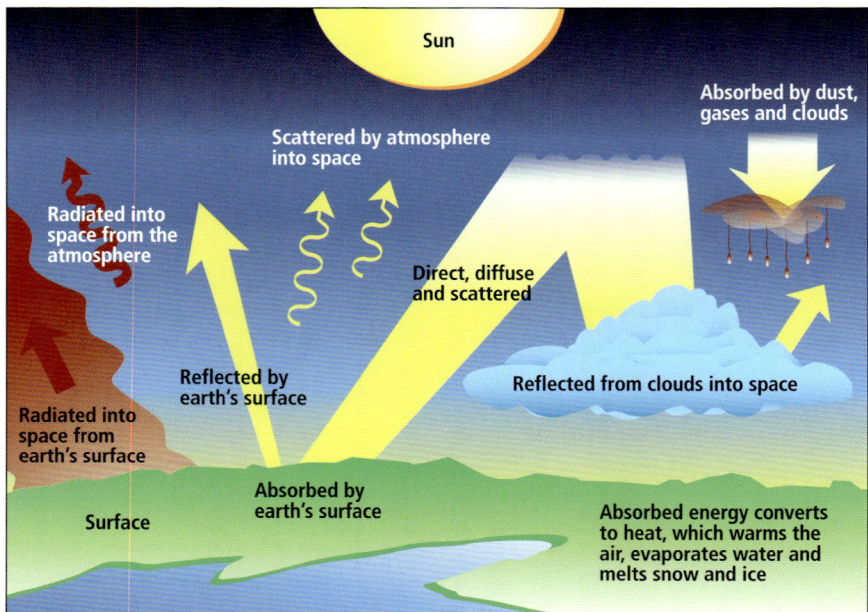

Fig. 1-4 Some solar energy is reflected and scattered back into space. The remainder is absorbed in the earth-atmosphere system, which radiates energy. Overall, it is almost in balance.

by moisture and clouds on the way out. There is a balancing act in constant play between the incoming and outgoing energy, and the average temperature of the earth has been reasonably steady over time. Indications of an imbalance in this energy account give rise to the global warming concept, which is discussed in a later section of this book.

The Ontario land surface, of course, plays a role in the global energy balance—the land surface is a mixture of cities, farmland, forest and lakes. As a result, the province is subject to wide variations in surface temperature just because of the surface characteristics.

Vertical Variation of the Atmosphere

As already noted, atmospheric pressure decreases with altitude, but what about other measures? Temperature is the most significant parameter. The behaviour of temperature is used to define the layers of the atmosphere. Rates of temperature change with height are known as thermal lapse rates.

At Ontario's latitude, temperatures in the lower atmosphere generally decrease from their highest values near the earth's surface to their lowest values at altitudes of about 10 to 12 kilometres. This layer of the atmosphere, characterized by

The Atmosphere

decreasing temperature, or negative lapse rate, is called the troposphere, and it is here that all active weather occurs. The warmth at the bottom of the troposphere is a result of the warmth of the land and water surfaces combined with the heat contained in water vapour. The temperature decrease stops abruptly or pauses at a level known as the tropopause, where temperatures are -50° to -60° C.

Moving upward in the atmosphere from the tropopause, the temperature slowly increases through the layer known as the stratosphere, until it reaches its maximum at a height of about 50 kilometres. The stratosphere contains most of the atmosphere's ozone, and warming through this layer is mostly a result of the absorption of the sun's ultraviolet radiation. The maximum temperature is often in excess of 0° C at this height. This maximum temperature level is known as the stratopause. Thereafter, temperatures decrease again through what is known as the mesosphere, approaching a minimum value near -90° C at the mesopause, which lies at a height of approximately 80 to 90 kilometres and has an atmospheric pressure of only about 0.01 mb. Above the mesopause, temperatures increase with height through a layer known as the thermosphere. In spite of low density within the thermosphere, the tenuous atmosphere can still affect drag on spacecraft at altitudes of 250 kilometres.

Moisture in the atmosphere plays a major role in the formation and development of weather. Virtually all the moisture is contained in the troposphere, mostly at the bottom of this atmospheric layer. The heating of the troposphere by the sun results in convective mixing, which distributes moisture and heat to higher levels. Gusty surface winds interacting with the terrain also mix this

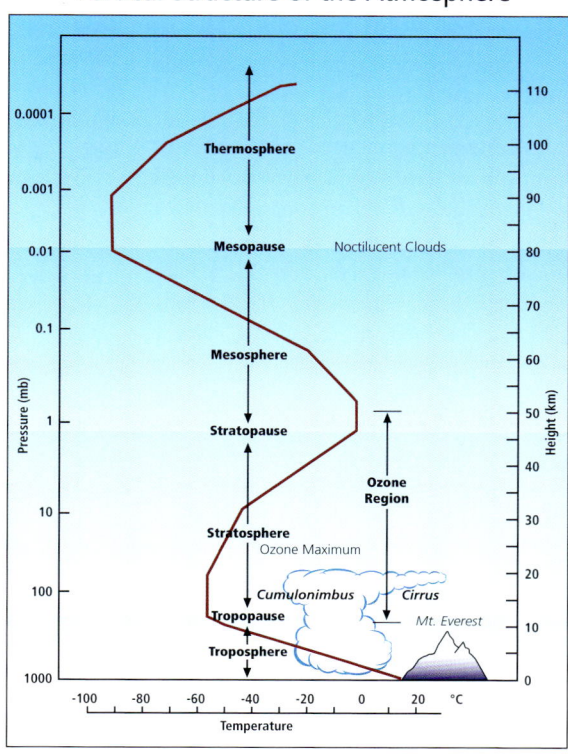

Fig. 1-5 Temperature changes with height and defines the different layers of the atmosphere. All active weather is in the troposphere.

15

The Atmosphere

moisture and heat upward by turbulent mixing. Moisture can also be transported to higher levels in the troposphere by the more gradual processes associated with what is known as frontal lift. Only rarely are these processes vigorous enough to drive moisture through the stable tropopause.

Atmospheric Circulation

Horizontal variation in atmospheric parameters has much more complexity than vertical variation. The earth's atmosphere can be thought of as a heat engine that is driven by the sun. In its simplest form, the basic function of the heat engine is to move heat from areas where it is most intense—in the equatorial zone—toward the cooler regions near the north and south poles.

When scientists started to think about atmospheric circulation, they developed simple models to explain apparently organized phenomena. George Hadley developed a simple model in 1735 to explain the elements of the easterly trade winds that are observed in the tropics. Hadley noted that the equatorward motion of air in the easterly trade winds must be balanced in some fashion to prevent the accumulation of mass in the equatorial zone. He postulated that upward motion and then poleward motion of air would result. However, the simple model he proposed does not fully explain the circulation and associated weather patterns observed around the globe.

A model of atmospheric circulation developed by Rossby in 1941 helps explain some of the complexity in the mechanism whereby heat is transported away from the equator toward the poles. This model, as depicted in Figure 1-6, has a considerable element of reality, especially for Ontario at so-called mid-latitudes. In this model, the Hadley circulation shows flow toward the equator, ascent and return flow aloft, and then subsidence in the subtropics. This is a direct cell, driven by heat. The subtropics, where air is descending, are regions of generally high pressure (or anticyclonic circulation) and sunny skies, which are well known to the "snow birds" who flee cold northern temperatures and head to those latitudes (from 20° to 30° N) for winter vacations. Another direct circulation cell is present at high latitudes (60° to 90° N). Here, cold air moves away from the pole toward the south. There is a general area of subsidence near the pole (surface anticyclone or high pressure), and a compensating northward flow aloft. Between the two direct cells is an indirect cell in which surface air is transported northward and air aloft southward. Warm, moist air from the south and cold, dry air from the north meet at the mid-latitudes—the zone where mid-latitude low pressure systems persist. This zone of air mass mixing, convergence and uplift associated with the mid-latitude low pressure systems provides all the key ingredients for production of the weather phenomena that we observe.

> *The trade winds get their name from the period in history when sailing ships were the main vehicles of trade. Captains of these ships factored in these winds when mapping out their journeys, sailing with the wind as much as possible.*

The Atmosphere

Hadley Circulation

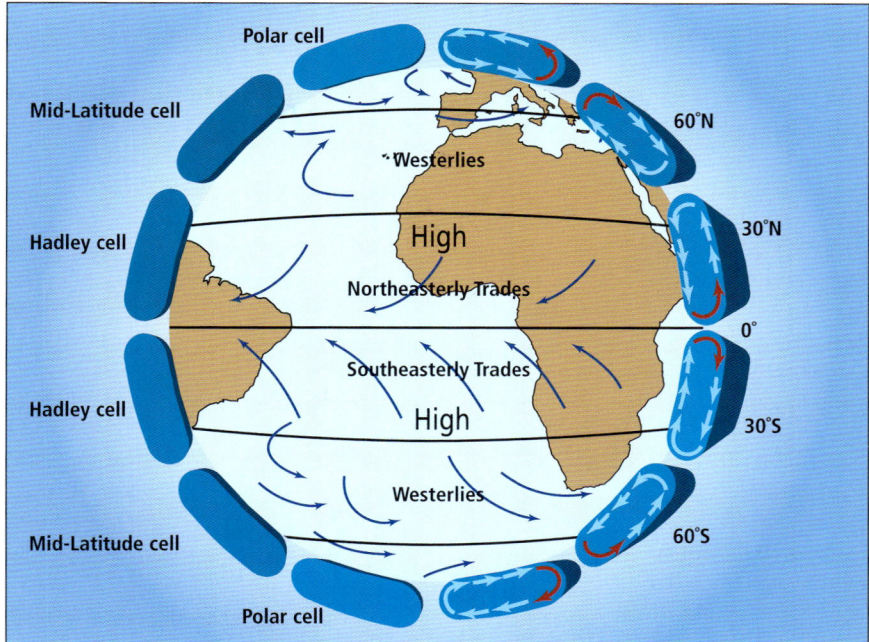

Fig. 1-6 A cross-section view of the atmosphere shows a three-cell circulation on each side of the equator. There is active weather in regions of general ascent (red arrows).

The simple circulation processes described do not take into account the rotation of the earth on its axis. This rotation has two important consequences. First, the zone of maximum heating moves from east to west as the earth rotates on its axis. Second, the earth's rotation also causes an apparent deflection of the airflow. On a rotating surface, an apparent force deflects any moving body. Gustave-Gasparde Coriolis first described this effect mathematically in 1835. North of the equator the deflection is to the right, and south of the equator the deflection is to the left. The magnitude of the apparent Coriolis force is proportional to the speed that the body moves.

Air parcels move because of the force of pressure. Two bodies of air that are separated horizontally with different temperatures have different densities. Essentially, the colder, higher density air mass has a higher pressure than the adjacent warm, less dense air. Pressure differences between air masses cause air

> Blow winds and crack your cheeks! Rage, blow,
> You cataracts and hurricanes, spout
> Till you have drenched our steeples, drowned the cocks!
> —Shakespeare, *King Lear*

The Atmosphere

parcels between the masses to move toward lower pressure. This activity is known as the horizontal pressure gradient force. And the greater the pressure difference (the larger the pressure gradient), the greater the force and therefore the faster the air parcels move. At the same time, the Coriolis force causes a deflection of the air parcels. If we ignore other forces, such as friction or the centrifugal acceleration in curved flow, the Coriolis force pretty well balances the pressure gradient force, and the air parcels move along lines of constant pressure, called isobars.

Representation of Weather Systems

The most convenient way to depict weather systems is by means of pressure patterns. Moving along a constant height surface, there is a reasonably large variation of pressure in the horizontal. For example, on a map drawn at sea level, pressure can vary by more than 100 mb. Atmospheric pressure is measured at many locations around the world. When pressures measured at the same time are plotted on a map, areas of high or low pressure can be discerned. An "L" marks the centre of an area with pressure lower than its surroundings, and an "H" marks the centre of an area of relatively high pressure. Lines joining locations that have the same pressure are called isobars. The greater the difference between the high and low pressure areas, the greater the number of isobars between those areas, and the greater the wind speed. So basically, a weather map is a depiction of pressure patterns with high and low pressure areas. In appearance, a weather map with isobars resembles a terrain map with elevation contours. A number of other parameters are measured at the same time as pressure, and these can also be plotted on the weather map. When the patterns of all these parameters are inspected, many relationships become apparent. Low pressure areas are generally associated with areas of cloud, and often with precipitation. High pressure areas are associated with relatively cloud-free areas. Warm temperatures are usually near low pressure centres, and relatively cool temperatures are in areas of high pressure.

Wind Deflection as a Result of Coriolis Effect

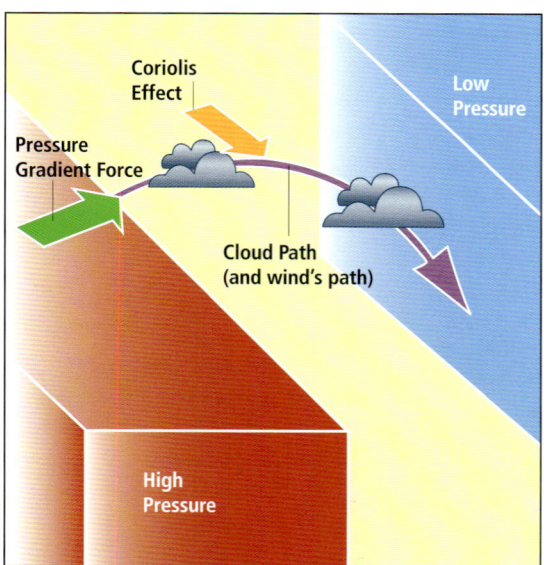

Fig. 1-7 Air moving from high pressure to low pressure is deflected to the right in the Northern Hemisphere.

18

The Atmosphere

Linkages Between Upper Air and Surface Patterns

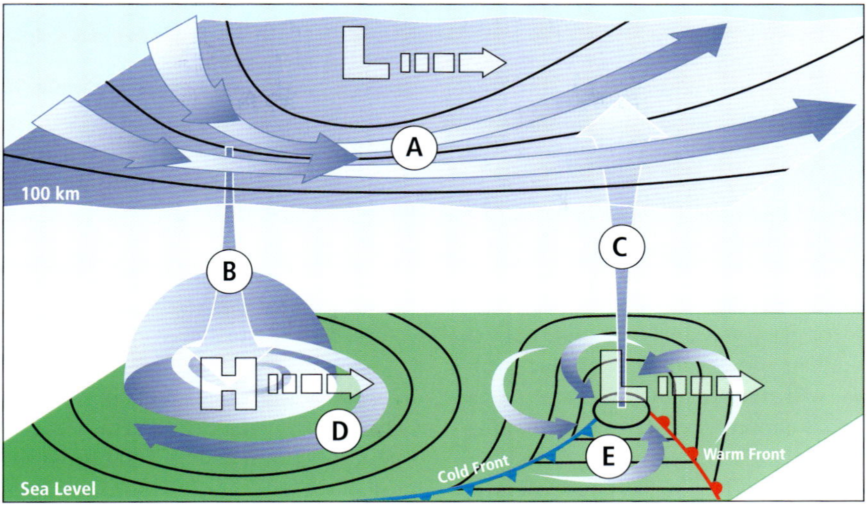

Fig. 1-8
A - wind flows around a moving low centre in the upper atmosphere
B - air subsides from an area of converging air aloft toward the ground
C - air ascends from a surface low toward an area of diverging air aloft
D - air spirals outward and clockwise around a surface high
E - air spirals inward and counter-clockwise around a surface low

Maps of conditions at other levels in the atmosphere can also be drawn. When depicting weather patterns at upper levels, the relationship between pressure and height is used. Rather than depicting pressure patterns at a constant height, constant pressure levels are chosen and the height of that pressure level above sea level is depicted. Meteorologists examine pressure patterns at several pressure levels within the troposphere. Again, patterns in this height field can be analyzed. As one progresses upward in the atmosphere, the patterns become less "busy" than those at the surface. There are fewer centres of high or low height than centres of high or low pressure on a surface weather map covering the same area. As well, the patterns of height appear to have a rather uniform shape, often with broad areas that have an undulating or wavelike appearance. The relationships between these patterns, their shapes, and the juxtaposition of patterns at different levels vertically then give the meteorologist a representation of the vertical structure of the atmosphere. Understanding the linkages between weather patterns through the depth of the troposphere is the essence of synoptic meteorology.

When there are four isobars off the Gulf of Mexico, expect rain within 24 hours across southern Ontario.

19

The Atmosphere

The study of synoptic meteorology has been underway since the early 1900s. When surface observations were combined with observations aloft and the patterns were analyzed, it became possible to develop a three-dimensional model of how the atmosphere was structured. The knowledge that the atmosphere has structure and behaves in an understandable way led to a number of theories about how weather patterns move and change with time. The science is complex and can be represented by the mathematics of fluid dynamics, which we will not attempt to reiterate—textbooks abound. But we can try to understand the elements of these atmospheric dynamics well enough to understand the formation, movement and prediction of weather systems over Ontario.

Two key concepts play a crucial role in understanding the behaviour of weather in Ontario. The first concept is that of air masses and the so-called frontal regions that separate air masses. The second concept is the link between the motion of weather systems and the atmospheric flow patterns aloft.

Air Masses and Fronts

Air that remains over a land or ocean surface for some period of time takes on characteristics dependent on the underlying surface. In meteorology, the term air mass is used to describe a large pool of air that has a uniform character. If the land surface is intensely heated by the sun, the overriding air will be warm, and if the underlying surface is snow or ice covered, the air will be cold. If the air is over an ocean, moisture will evaporate into the lower portion of the air mass.

Air that remains over central regions of North America during summer is called a continental polar air mass, and we frequently experience this air mass. When the circulation is more active, air masses stream into Ontario from other regions. Air that originates from the northern Pacific Ocean, known as maritime arctic air, often crosses Ontario. Air from the arctic regions, which is dry and usually cold, is known as a continental arctic air mass. In winter, this air mass gives us the coldest (occasionally record-breaking cold) conditions. Maritime tropical air masses, originating over low latitude regions of the Atlantic or Pacific Oceans, frequent Ontario and are responsible for most of the large precipitation events.

The boundaries between adjacent air masses are known as fronts. Between World War I (1914–18) and World War II (1939–45), Norwegian meteorologists established the frontal concept, as it is known, to give some structure to weather systems that influence mid and high latitudes. This concept is a key building block of modern meteorology. Weather conditions are usually described in relation to the presence, motion and modification of frontal zones. When air from a warm air mass moves against a retreating cooler air mass, the boundary is known as a warm front. When the colder air mass prevails and advances against the warmer air, we have a cold front. As

> *Lows bring strong winds, clouds and rain. Highs bring clear skies, perhaps with some high cloud, and can result in very cold weather in winter.*

The Atmosphere

Fronts

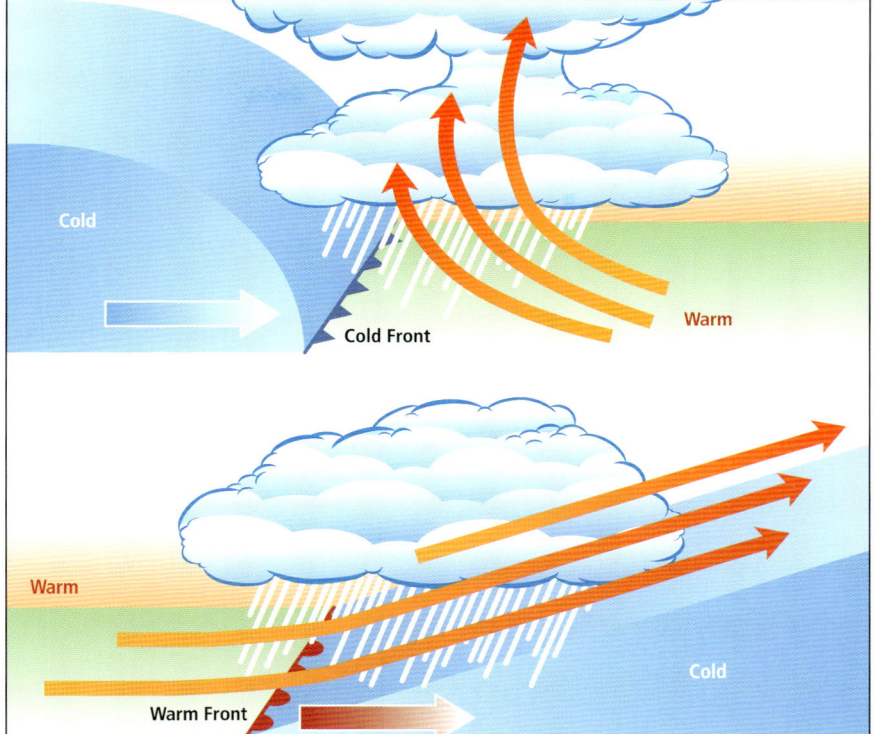

Fig. 1-9 With a cold front, warm air is displaced by advancing cold, dense air and clouds are formed. With a warm front, warm air overruns the retreating cold air, forming clouds and precipitation.

at other locations in the mid-latitudes, Ontario's weather is frequently characterized by the approach and passage of warm and cold fronts.

Another important aspect of the frontal model of the atmosphere is that regions of low pressure frequently form along the frontal boundaries between air masses. The characteristics, behaviour and motion of these low pressure systems are the key to describing present and future weather conditions.

Motion of Weather Systems

The development and movement of frontal storms and weather systems is tied to weather patterns in the mid to upper troposphere. The patterns differ between the northern and southern hemispheres, and there is little apparent interaction between patterns across the equator. As shown in Figure 1-10, the wind flow at these upper levels usually displays a wavy, undulating pattern. Meteorologists characterize these patterns as either long waves or short waves.

The Atmosphere

Long waves are known as planetary waves because they are discernable over the whole planet. The planetary wave, with its characteristic trough and ridge pattern, is continuous around the hemisphere. Over a period of days to weeks, there may be an apparently stable configuration in the planetary pattern with a distance of several thousand kilometres between successive troughs. These planetary waves usually move slowly eastward at a speed of about 10 to 15 kilometres per hour. Planetary waves will, on many occasions, appear to remain stationary, and portions of the pattern may even move westward, a rare phenomenon known as retrogression.

Short waves, on the other hand, move relatively quickly. They appear as ripples in the planetary wave pattern and move along the long wave pattern from west to east. The speed of these waves can be fairly high, often 40 kilometres per hour or more. Short waves are closely tied to the so-called synoptic scale storms that are 300 to 500 kilometres across. In summary, short waves, with their attendant storm systems, move along, or are steered by, the planetary wave pattern.

The speed of the wind itself within the steering flow is usually much greater than the rate of progression of the short

> *The jet stream was not discovered until World War II, when pilots noticed they lost ground speed when flying against it.*

This image reveals a lot of wind throughout the depth of the atmosphere. Stratus fractus cloud is shredded by turbulent mixing and strong winds in the low level. Mid- and high-level clouds are all shaped by the winds at those levels. The clouds are like a book waiting to be read.

The Atmosphere

Wave Pattern in Wind Flow

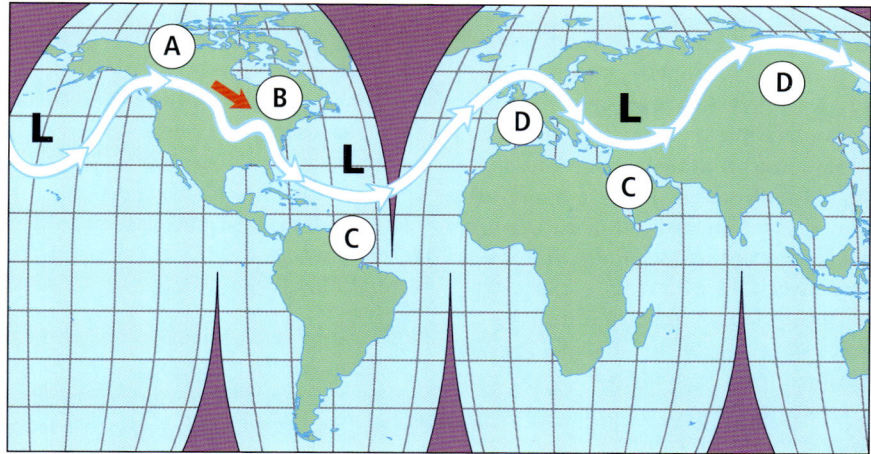

Fig. 1-10
A - jet stream
B - short wave moving through ridge
C - trough in planetary wave
D - ridge in planetary wave

and planetary waves. As well, the amplitude of the short wave ripples usually decreases the higher we go in the atmosphere. The short waves are most apparent at the middle level of the troposphere; therefore, meteorologists pay a lot of attention to their shape, rate of motion and other characteristics. The long wave pattern is most evident at the top of the troposphere, and winds are generally strongest at this altitude. These winds are known as the jet stream. Meteorologists often designate the jet stream as the steering flow for short waves and storms.

The location of the ridges and troughs in the planetary wave pattern also plays a role in determining average weather conditions. When the pattern is relatively stable, the associated weather conditions are also stable. Under a ridge, an area of high pressure dominates at the surface. High pressure areas are associated with subsiding, or a downdrift, of air. There is usually little cloud, and fair skies with sunny conditions are prevalent. On the other hand, a trough is associated with unsettled weather. When short waves round the bottom of the planetary trough, the associated weather systems become quite active. A considerable amount of vertical motion (ascending air) may be generated through the mid-troposphere, usually producing cloud, stormy conditions and precipitation. The distance between the trough and ridge in the planetary wave pattern is typically 2000 to 3000 kilometres. When the pattern is stable over Canada with the ridge positioned over the west, the trough is positioned over the east. As a result, warm, dry conditions will prevail over the Rockies and western prairies while unsettled and perhaps cooler conditions will be prevalent over Ontario. Of course, the reverse situation also occurs.

The Atmosphere

Weather Influences in Ontario

Ontario is located within the mid-latitude zones where air masses always vie for supremacy. Tropical air from the Gulf of Mexico often emerges from the south. Arctic air masses from the tundra are always nearby, to the north. Continental air masses from the Great Plains of North America are sandwiched in between. To make things exciting, the steering flow and associated jet stream are also typically in close proximity, directing the motions of these air masses. Variable weather must surely result as the contrasting air masses and the associated fronts sweep back and forth over the Ontario landscape.

Most of the strong low pressure areas affecting Ontario develop as a result of the Rocky Mountains. As a slab of air that extends from sea level to the tropopause approaches the west coast of North America and begins to traverse the mountains, the depth of the air mass decreases—the bottom is pushed up. Winds curving around the base of a low pressure system decelerate on the upward, windward slope and accelerate on the leeward slope. Effectively, a local ridge of high pressure is observed over the mountain range. As the air mass travels to the east of the Rockies, it stretches as it moves down the descending land surface. Low pressure systems may intensify as they move into the prairies east of the Rockies, or a new surface low pressure system may form. This process is known as lee cyclogenesis, and it happens whenever westerly winds are forced to flow over a mountain range.

As a result, the strongest lows affecting Ontario typically originate from the lee of the Rockies. Colorado lows tracking northeastward toward Ontario typically bring heat, moisture and a lot of precipitation from the Gulf of Mexico. Lows born in the lee of the Rockies over Alberta are referred to as Alberta Clippers and can bring significant precipitation followed by a cold snap as they race southeastward across Ontario.

These air masses and systems are also modified by the landscape. The Great Lakes are more like inland seas than lakes. In winter, heat and moisture from the lakes are added to the overlying air masses, potentially supercharging low pressure systems. November storms are remembered by landlubbers and mariners alike for their hurricane force winds and heavy precipitation of all types. In summer, cooling from the surface of the lakes actually weakens low pressure areas while building high pressure areas. The Niagara Escarpment, Algonquin Highlands and James Bay Lowlands all play a role in altering air masses and weather as they arrive.

Some Ontario weather is homegrown. Winter snowsqualls off the Great Lakes are vital to the ski community, and the gentle slope of the James Bay Lowlands often makes Moosonee the surprise hotspot in Ontario during summer.

These factors—the proximity to diverse air masses, the undulating storm steering flow, the waters of the inland seas and the countless other terrain features—combine to make weather very interesting in Ontario and very challenging to predict.

"Windy Cumulus" by Phil Chadwick

Chapter 2: Clouds

> The cloud never comes from the quarter of the horizon from which we watch for it.
> —Elizabeth Gaskell

For many people, clouds are the most interesting aspect of weather. Our perception of the coming weather for each day is based on our first glimpse of the sky and the presence or absence of clouds.

Clouds consist of tiny water droplets or ice crystals that occur as a result of the condensation of atmospheric water vapour. Condensation occurs with great difficulty in clean air, so atmospheric water vapour must have a surface upon which it can condense. The formation of cloud particles is therefore dependent on the availability of tiny particles known as condensation nuclei. These condensation nuclei are almost always abundant in the atmosphere, and they arise from dust, pollution particulates, pollen and spores that are stirred into the atmosphere as a result of wind motion and turbulence. In most regions of the lower troposphere, the density is at least 10,000 per cubic centimetre. In industrial areas, concentrations can increase to more than one million nuclei per cubic centimetre. There is no shortage of condensation nuclei available for condensation!

25

Clouds

At temperatures above freezing, the cloud droplets that form on the condensation nuclei are composed entirely of water. At temperatures between freezing and -4° C, the cloud droplets are more likely to be comprised of supercooled water droplets because most condensation nuclei do not encourage the growth of ice crystals until they are colder than -4° C. The probability that the clouds are composed of ice crystals increases as temperatures fall. Supercooled water droplets in a cloud or fog can have enormous impacts on transportation and safety.

There is an upper limit on how much water vapour can be present in the free atmosphere. The limit depends on the temperature and pressure of the air containing the moisture. Relative humidity is the main way we express water vapour content in the atmosphere. The relative humidity is a ratio of the actual amount of water vapour in the air compared to the maximum amount of water vapour that the air parcel could contain at the same pressure and temperature. This relative humidity is expressed as a percentage—air that is saturated with water vapour has a relative humidity of 100 percent. So, when clouds are formed, the relative humidity of the air containing the cloud must be 100 percent.

Another term that is often used to measure atmospheric moisture is the dew point temperature. When unsaturated air is cooled at a constant pressure and without the addition of moisture, the temperature at which the air becomes saturated and dew starts to form is referred to as the dew point. The dew point temperature is a value that is often representative of the widespread characteristics of an air mass.

For cloud to form, there must be a mechanism for the air to increase its relative humidity to 100 percent. When air that contains water vapour cools, the vapour transforms into cloud droplets. Occasionally, fog forms when a moist air mass moves over a colder surface and cools from below. Generally, however, air cools and expands as it ascends, moving into areas of lower pressure. The rate of cooling of the air parcel in the lower half of the atmosphere is approximately 10° C per kilometre of ascent. Usually the water vapour remains within the parcel of air and cools at the same rate. When the water vapour cools to a temperature at which condensation starts occurring on condensation nuclei, the air parcel is described as saturated. Its relative humidity is 100 percent, and cloud forms. This altitude is then the base of the cloud. When the air parcel is lifted from the earth's surface, this altitude of cloud formation is typically referred to as the lifted condensation level of the air mass.

If the parcel continues to rise, the rate of cooling changes. In fact, the condensing water vapour releases latent heat that was stored during the formation of water vapour by evaporation. This release of latent heat effectively cuts the rate of parcel cooling by about half, which has important consequences for development of convective clouds and thunderstorms.

Ascent of air can occur in a number of different ways. The sharpness of the cloud edges can be used to infer the intensity of the forces creating the clouds.

Orographic Lift

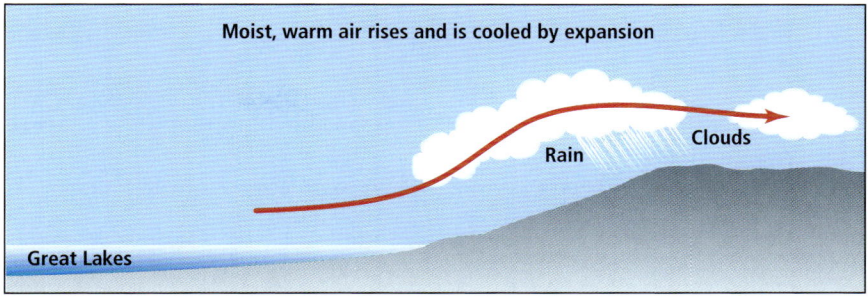

Fig. 2-1 Orographic lift occurs when winds blow uphill. If the air is moist, cloud—even precipitation—may form. Because of the prevailing westerlies, western slopes are more likely to be cloudy while the eastern slopes will have less cloud due to orographic descent.

Dynamic Lift

This lifting process results from the motion of the atmosphere at middle to upper levels in the troposphere. It is associated with the frontal lifting process in that the ascent results from the relative motion of air masses. The ascent may be more widespread and gentle than that associated with the immediate frontal surfaces. Generally, broad scale ascent is associated with the counter-clockwise motion around a low pressure system. Descent of air, or subsidence, is associated with clockwise flow around a high pressure system.

Orographic Lift

Air can be forced to ascend as it moves along sloping terrain in a process known as orographic lift. Because of the presence of highlands, escarpments and sloping plains, this process is very relevant in Ontario. The influence is predominant in the lower atmosphere because Ontario lacks anything even approaching the height of the Rocky Mountains, which can extend orographic lift well into the upper troposphere.

Frontal Lift

A warmer air mass must ascend over a colder, denser air mass. The interface between the two air masses is known as the frontal surface. One process that often governs ascent of air is associated with these atmospheric frontal surfaces. This process is in many ways similar to orographic lift, except that the lifting agent is the frontal surface that surrounds the colder air rather than the earth's surface. Frontal surfaces can extend through the entire troposphere, so frontal lift can occur through a considerable depth.

Cold frontal lift occurs when the cold air is advancing and produces a surface slope of one unit of rise for every 100 horizontal units. Warm frontal lift occurs when cold air is retreating and produces a shallow frontal slope about half as steep as that associated with advancing cold air and a cold front.

Frontal type lift also occurs when the warmer air mass is lifted completely above the surface by a layer of surface-based cold air. Typically, the warm air is

Clouds

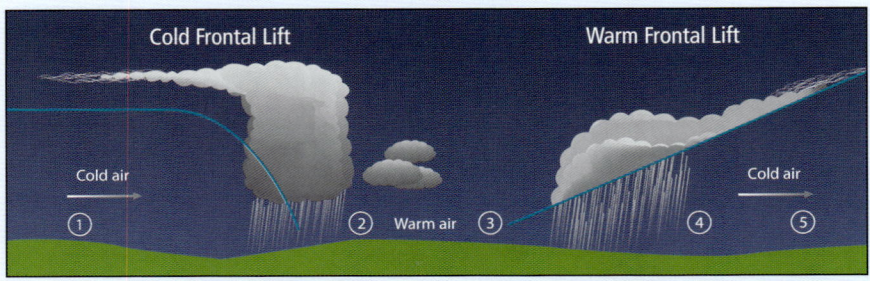

forced to follow an elevated valley formed within the surface-based colder air. This trowal (trough of warm air aloft) structure is a Canadian concept that was first developed in the late 1940s. The common trowal typically flows northward from the apex of the warm and cold front connection. The warm air from the warm sector of a low pressure area follows the elevated valley in the cold surface air and rises as it heads typically northward. The frontal type lift aloft in the trowal can produce cloud and precipitation at great distances north of the surface fronts. There is no change of air mass at the surface as the main system passes well to the south.

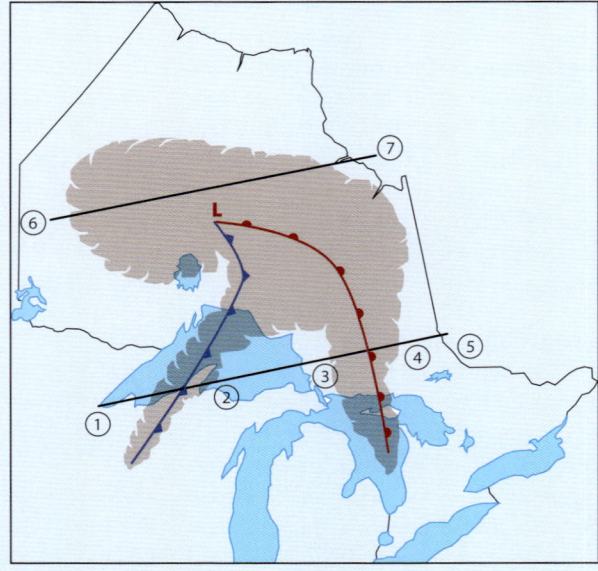

Fig. 2-2

28

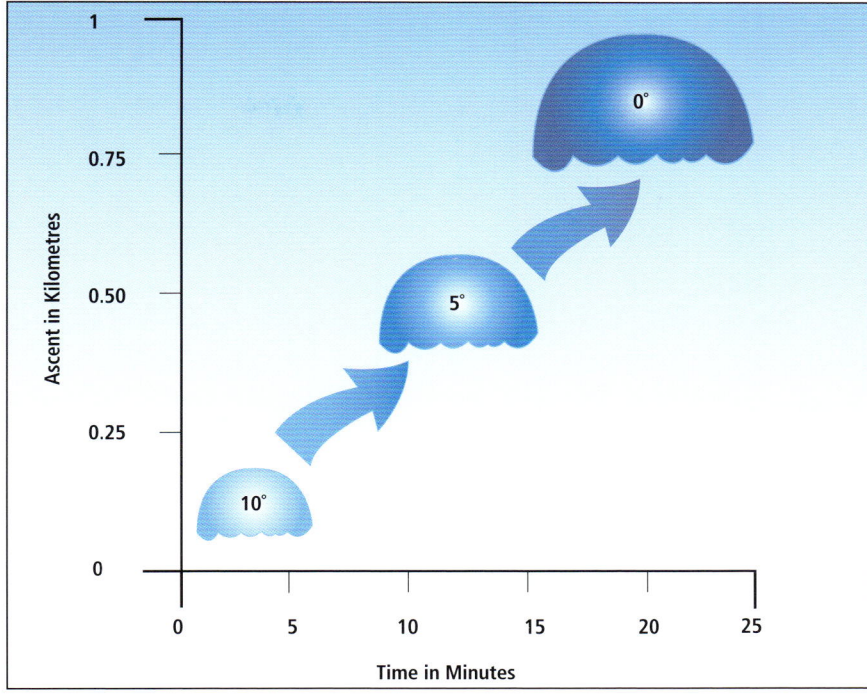

Fig. 2-3 Parcels of air rising as a result of convection

Convection

Another important mechanism that produces ascent is convection. This process is common during daylight. As the sun heats the land surface, air adjacent to the land warms. The amount of warming depends on the type of land surface and its aspect ratio to the incident rays of the sun. Some air parcels become a little warmer than adjacent parcels, and they rise buoyantly. As they rise, they cool. If these parcels remain warmer than the adjacent air at that level, they continue to rise. This process continues until the rising air parcels reach a level at which the surrounding air mass is warmer. At that level, the upward buoyancy ceases and the parcel spends its kinetic energy by initially overshooting this equilibrium level to which it will eventually return. Hot air balloons rise through the same process.

Turbulent Lift

This mechanism is caused by the friction between a moving air mass and the underlying surface. Turbulent lift is somewhat similar to convection except that air parcels get their upward motion from undulations in the underlying surface. The height to which the turbulence extends depends on the speed of airflow (i.e., wind speed), the roughness of the underlying surface and the stability of the air mass itself. Turbulent lift generally does not extend far above the earth's surface, but it can be horizontally widespread.

Clouds

Cumulus clouds along the St. Lawrence River near Brockville

Cloud Classification

An international classification system for clouds has been developed for use by ground observers. Clouds are generally classified into four families based on how high the cloud base is located above ground level. Low clouds have bases in the layer that reaches up to approximately 2 kilometres above ground level, though cloud based at ground level is generally called fog. Middle clouds are based between about 2 and 6 kilometres above ground level and are known as alto form (Latin for "middle"); high clouds are based above 6 kilometres. Clouds that have a considerable vertical dimension and that are based in either the low or middle range are called cumulus (Latin for "heap").

Each cloud family is subdivided based on appearance. In general, the visual appearance of clouds is closely associated with the process that gives rise to the lifting process. Clouds that appear as a layer and are spread out horizontally are called stratus clouds (Latin for "layer"). Most high clouds are wispy in appearance, largely because they consist predominantly of ice crystals that are formed at the cold temperatures associated with the upper troposphere. These clouds are known as cirrus (Latin for "curl of hair"). Latin terms are used to further define appearance (fractus, castellanus, pileus, lenticular, mammatus) or other cloud characteristics, such as the presence of rain. These terms are often used in combination. The nimbus subdivision is used for cloud types from which precipitation is falling.

What follows is a detailed description of the cloud types seen in Ontario.

Stratus

As observed from the ground, stratus clouds appear as a uniform layer usually resulting in an overcast condition—that is, the sky is uniformly covered from horizon to horizon. These clouds are usually made up of liquid water droplets, but if the temperature at cloud level is between freezing and -4° C, stratus clouds are probably composed of supercooled droplets that freeze on contact. At temperatures colder than -10° C, these clouds become increasingly likely to be composed of ice crystals. Usually these clouds do not have any obvious markings on their lower surface, and individual cloud elements are indistinct. Precipitation from stratus clouds is rare and is at most a drizzle, freezing drizzle or light snow. You usually can't see the sun through stratus clouds, but if you can, the solar disk has sharp edges.

Stratus Fractus

These clouds have a ragged appearance and are based at the same level in the atmosphere as a stratus deck. Stratus fractus clouds result when the conditions that formed a stratus cloud deck are no longer active and the stratus deck breaks up.

Clouds — Low Clouds

Stratocumulus

Stratocumulus clouds consist of a series of patches of rounded clouds that have relatively little vertical development. These clouds are often associated with stratus clouds and are either a precursor to the formation of stratus or result from the transition of a stratus deck into cumulus cloud. The deck of stratocumulus clouds is not continuous; the edges are ill defined, and the cloud has a patchy or rolling appearance. Blue sky or higher cloud decks frequently appear through breaks in the layer of stratocumulus. Any precipitation associated with these clouds is at most a drizzle or light snow and requires a cloud thickness in excess of 500 metres.

Fog

Fog can be considered a cloud that is in contact with the ground. It can form when a moist air mass flows over a colder surface, or in calm wind conditions when moist air in contact with the ground cools overnight to the point that its relative humidity rises to 100 percent. When fog breaks up, usually in the presence of a slight breeze, a layer of stratus and eventually stratocumulus clouds typically forms above the ground. As ground heating increases in conjunction with an increase in surface wind, the base of the stratus deck may rise. Fog may also dissipate because of the action of sunlight, which heats the top of the fog layer. Some of the solar energy also penetrates the fog deck, warming the underlying ground surface and raising the air temperature beyond the point of saturation. Heating of the ground on the edge of a fog bank also initiates air circulation that eats away the cloud from the outer edges inward. Technically, to be considered fog, the cloud must reduce the visibility to less than 1 kilometre.

Clouds — Low Clouds

Radiational Fog
Created by cooling from below, this type of fog results in locally poor visibility. Radiational fog is most common in autumn when nights are getting longer and cooler at the same time that vegetation is drying out and providing moisture to the air. It forms in valleys and especially over water surfaces.

Mist
Mist is a ground-based obstruction to visibility through which one can still see 1 kilometre or more. Precipitation adding moisture to an air parcel is also likely to result in mist. Warm rain creates the mist in the accompanying image.

Low Clouds — Clouds

Advection Fog
This type of fog is created when warm, moist air flows over a cold surface, such as ice or water. The air parcels are brought to saturation by cooling from below. Advection fog is most common in spring when the water is still frigid but warm, moist air masses are starting to push up over the Great Lakes.

Arctic Sea Smoke or Steam Fog
Caused when stable, frigid arctic air flows over warm water, this fog is heated from below and rises buoyantly. Arctic sea smoke is most common in late autumn and throughout winter over open water.

35

Clouds — Low Clouds

Nimbostratus

As the nimbo preface implies, this cloud is often associated with continuous, widespread rain or snow. This cloud is generally quite thick, extending from 1 to 2 kilometres above the ground at its base to tops often over 5 kilometres. The cloud base is usually dark grey to black, and cloud elements are indistinct. Nimbostratus must be a minimum of 1000 metres in depth to successfully produce precipitation.

Low Clouds — Clouds

Turbulent stratocumulus

Cloud Streets

Low clouds align along the direction of low-level wind, resulting in what is known as a cloud street. These clouds are generally stratocumulus or cumulus. A steady horizontal wind initiates a rolling motion in the flow. Bands of gentle ascent separate bands of gentle descent, and both parallel the average wind direction. A line of clouds forms in the areas of ascent.

Turbulent stratocumulus clouds, which form when strong winds turbulently lift the air to saturation, are common in Ontario. Rows or streets of turbulent stratocumulus typically form parallel to the average wind direction in the lowest layers of the atmosphere.

Helmholtz Waves

Helmholtz waves resemble breaking waves on water. These clouds are formed along a horizontal interface between warm air that is flowing over colder air. They are rarely seen because there is not usually enough moisture at that interface for clouds to form.

Clouds — Middle Clouds

Altostratus

Altostratus has a grey to steely blue colour and generally covers the whole sky. This cloud has no characteristic thickness and can be up to 6 kilometres deep, which is thick enough to completely obscure the sun. If you can see the sun through altostratus, it looks much as it would if you were looking at it through frosted glass. The base of the cloud appears flat when observed from ground level, but if you are flying through an altostratus layer, the lower boundary is virtually indistinct. Precipitation rarely falls from these clouds, but precipitation will develop if altocumulus continues to thicken to transform into nimbostratus.

Middle Clouds — Clouds

Altocumulus

This cloud type is similar to stratocumulus. The altocumulus cloud deck is not usually widespread. The cloud elements are arranged in groups. Within the groups, the elements frequently form into lines that are perpendicular to the wind at the cloud layer in much the same way as water waves are perpendicular to the wind on the surface of a lake.

Clouds — Middle Clouds

Altocumulus Castellanus

These clouds usually develop within a line of altocumulus. They are an indication that the middle atmosphere is unstable. When they form in the mornings during summer, they occasionally grow to be quite large vertically, and precipitation can fall from them. Occasionally they even develop into thunderstorms.

Clouds — Middle Clouds

Lenticular and Wave Clouds

These clouds are often observed in the vicinity of hills and mountain ranges. When air that is thermally stable vertically is forced to flow over obstructing terrain, it frequently forms a standing wave. Standing waves along barriers are well known to glider pilots, who use them to provide the lift necessary to soar to heights above 10 kilometres. The moisture flowing through the wave pattern condenses in the updraft portion of the wave. Where the air subsides, cloud dissipates, giving the cloud the form of a curved lens. Because the waveform is linked to the obstructing terrain, the cloud is usually stationary in the sky for as long as the upper level winds and moisture conditions are favourable.

Lee Wave Wind Pattern with Lenticular and Rotor Clouds

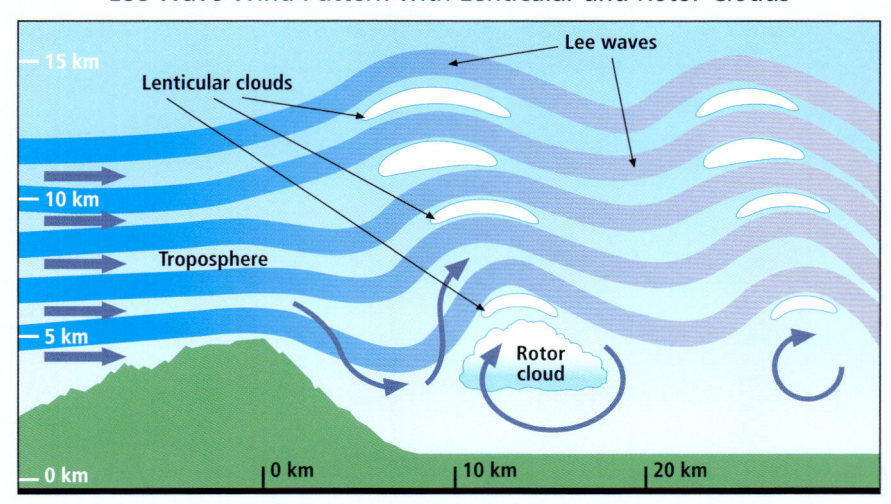

Fig. 2-4 A cross-section of wind/flow associated with standing wave pattern

Clouds — High Clouds

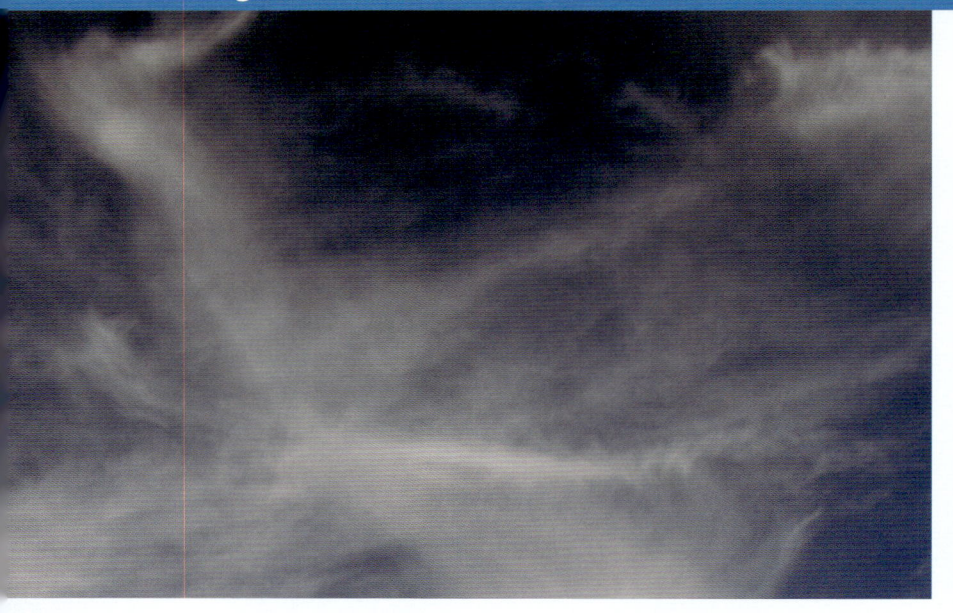

Cirrus

Cirrus clouds consist entirely of ice crystals. These clouds can take on a variety of forms but generally appear as fibrous wisps showing white against the blue sky (above). Isolated tufts occasionally have feather-like plumes that stretch out and turn upward. They are often called mares' tails (below).

High Clouds — Clouds

Cirrostratus

Also made up of ice crystals, cirrostratus clouds appear as a thin, whitish veil that can extend from horizon to horizon, giving the sky a milky appearance. Occasionally there is a fibrous appearance to the cirrostratus deck, which implies that there are streaks of thicker strands of cirrus cloud embedded within the deck. Cirrostratus clouds are thin enough that you can see the sun distinctly, as though you were looking at it through a thin veil. Halos are sometimes seen around the sun or the moon as it shines through the cirrostratus deck. The halo diameter is larger than the size of an outstretched hand at arm's length.

A cirrostratus deck is formed from the overriding of moist air along an elevated frontal surface. This indicates the approach of a different air mass typically associated with a low pressure area. More cloud and possibly precipitation is approaching.

Clouds — High Clouds

Cirrocumulus
These clouds have a patchy or, occasionally, wave-like appearance.

High Clouds — Clouds

Noctilucent Cloud

Noctilucent clouds occur in the mesopause at an altitude of about 80 kilometres. These clouds are made up of ice crystals, and the condensation nuclei are thought to be meteoric dust particles. The moisture content at this level is patchy, so the clouds are rarely observed. Because these ice crystal clouds are thin, they are only observed while the sun is still below the horizon. They have no effect on weather and are more of an observational curiosity.

THE COLOUR OF CLOUDS

You can often tell what is going on inside a cloud by what colour it is. Storm clouds are dark because they have a high water droplet content and the droplets are tightly packed together, so light has trouble passing through them. Thunderstorm clouds take on a green tinge because light is scattered by the ice in the cloud. If you see a green-tinged cloud, you can be pretty sure that heavy rain and hail are on the way. A yellow cloud gets its colour from the presence of smoke, usually from forest fires. Yellow clouds are rare but can sometimes be seen during late spring or early autumn, when forest fires are most common.

Clouds — High Clouds

Contrails

Aircraft exhaust contains large amounts of water vapour and particulate matter. The exhaust gas cools quickly and, with the cold temperatures at flight level, usually results in the formation of ice crystal clouds that appear to stream out behind aircraft. The turbulent motion of the atmosphere causes these contrails to spread laterally, and the turbulence created by the aircraft means contrail dissipation usually occurs within a few minutes. However, if the atmosphere at the contrail level has a high relative humidity, the contrails may persist for a while longer. The horizontal motion of the atmosphere at aircraft flight level further complicates the phenomenon. While successive aircraft fly along the same path relative to the ground, the air at flight level is almost always moving. This gives the impression that successive contrails are side by side. The presence of successive and perhaps long-lived contrails may result in the formation of a thin cirrus shield. This shield will grow throughout the day, and in areas that are frequently crossed by over-flying aircraft, a fairly extensive deck of cirrostratus can result. The jets using Toronto Pearson International Airport, Canada's busiest airport, provide an abundance of contrails.

Vertical Development — Clouds

Clouds with Vertical Development

The temperature of the ambient air that a parcel of rising air encounters governs how high a cloud can grow. If the temperature of the air within the parcel is warmer than the surrounding ambient air, the parcel continues to rise. When the air contained within a rising parcel becomes saturated, the latent heat energy is released. The cooling rate of the saturated air is reduced to about one-half of the cooling rate of the unsaturated parcel. Because the surrounding ambient atmosphere generally cools at a rate between the moist and dry rates, the rising saturated air parcel remains warmer than the surrounding ambient air. In this situation, the atmosphere is said to be convectively unstable, and clouds experience rapid vertical growth.

Cumulus

These clouds generally have a flat base at the lifted condensation level, and they are as high as they are wide. The edges of the clouds are usually quite distinct, giving the clouds the puffy, white appearance of popcorn. These fair weather cumulus clouds, often known as cumulus humilis, generally go through a standard life cycle. In late morning, the intensity of solar heating reaches a point where parcels of air near the surface rise, forming puffy cumulus clouds (we say the "cumulus starts popping"). As the day progresses, the appearance of the individual clouds slowly changes, and they may seem to be relatively motionless in the sky. Late in the day, the sun intensity decreases and the clouds generally dissipate. Precipitation rarely falls out of these cumulus clouds.

> *You can identify clouds by using your hand held out at arm's length:*
> - *Cumulus clouds are the size of your fist, or larger*
> - *Altocumulus clouds are about as wide as one or two fingers*
> - *High-level cirrocumulus clouds are about as wide as your pinky finger or smaller.*

Clouds — Vertical Development

Towering Cumulus

The cumulus cloud base may be 1 to 2 kilometres above the ground, and the tops of towering cumulus may extend to heights of 4 to 5 kilometres. These clouds contain rather vigorous updrafts and often produce precipitation in the form of rain showers or snow flurries—the precipitation starts and ends abruptly, much like the water in your shower. Towering cumulus clouds may become more organized and can grow to a significant extent horizontally as well as vertically. These larger groups of towering cumulus are known as cumulus congestus.

Vertical Development — Clouds

Occasionally, the tops of towering cumulus will give an upward push to a layer of moist air aloft. This moist air condenses, forming a veil-like layer of ice crystal cloud (essentially cirrus). These veiled cumulus congestus clouds are known as cumulus pileus.

TYPE OF CONDITIONS CLOUDS USUALLY BRING:

cirrus — *usually means fine weather*
cirrocumulus — *the mackerel sky is sometimes an indication of unsettled weather*
cirrostratus — *approaching rain or snow*
altostratus — *rain or snow likely if cloud thickens*
nimbostratus — *rain or snow*
stratus — *maybe light rain or drizzle*
cumulus — *sunny days*
cumulonimbus — *thunderstorms, showers and sometimes hail*

Clouds — Vertical Development

Typical Tornado-Producing cumulonimbus (CB) Cloud
...and associated features...

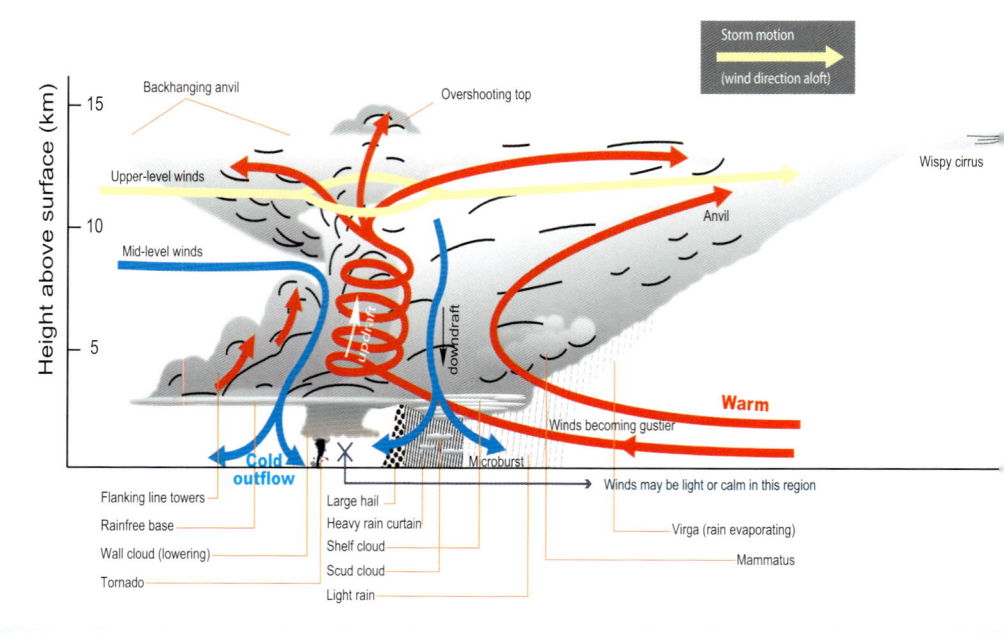

Fig. 2-5

Cumulonimbus

The cumulonimbus cloud represents the ultimate growth phase of the cumulus cloud into a vigorous, precipitating cloud mass. Cumulonimbus clouds can stretch horizontally for kilometres and extend from their base through the entire depth of the troposphere, to heights of 10 to 15 kilometres. Lower portions of these clouds contain liquid moisture (cloud and rain droplets); upper portions consist of ice crystals. The upward growth of these clouds is constrained by the thermal stability of a stable layer, which is typically the tropopause. The tropopause acts as a lid, and the tops of cumulonimbus then spread laterally, forming a fibrous cirrus veil. Strong winds aloft generally move the veil downwind, resulting in the classic anvil-like appearance of the cumulonimbus cloud. The anvil top can extend many kilometres downwind of the main cloud base. A weak cumulonimbus cloud is guided by the winds, whereas a strong cumulonimbus cloud evolves and propagates at an angle to the wind. This cloud almost always generates showery precipitation. More vigorous storms produce hail within the cloud, and this hail frequently reaches the ground.

Vertical Development — Clouds

Cumulonimbus clouds come in three distinct forms. They may appear chaotic, but they are actually orderly machines that ingest warm, moist air and convert it to energy of motion and potentially heavy precipitation and lightning.

Pulse Type Cumulonimbus

The first updraft resulting from daytime heating and some lifting mechanism is called a cell. If the wind does not change in speed or direction with height, this cell will grow vertically until it gets high enough for precipitation and lightning. The precipitation falls and cools the surface that originated the cell, effectively extinguishing the cumulonimbus cloud. This is a pulse type thunderstorm that typically completes its life cycle in less than an hour. In all but the most unstable atmospheric conditions, a pulse type thunderstorm brings precipitation but no severe, damaging conditions.

Clouds — Vertical Development

Supercell cumulonimbus near Chatham on July 18, 2007

Multicell Cumulonimbus

If the wind changes enough in speed and/or direction so that the downdraft associated with the precipitation should fall a distance from the updraft, a multicell cumulonimbus results. The first cell that caused precipitation is replaced by another cell that tends to form on the right-rear flank (looking in the direction that the thunderstorm is moving). This cell proceeds through its life cycle only to be replaced by another cell that again typically forms just southwest of its predecessor. The result is that multiple cells cycle through the cumulonimbus, which continues to survive through regeneration. Multicell cumulonimbus clouds are very common and last for one or two hours. They can produce heavy rain, damaging downbursts, large hail and, infrequently, short track tornadoes.

Supercell Cumulonimbus

If the environmental wind velocity veers (turns clockwise) at a rate of 20 to 30 kilometres per hour per vertical kilometre of lift, the updraft of the first cell will likely begin to rotate. Cumulonimbus clouds with a rotating updraft are called supercells, and they are responsible for all types of severe conditions and most of the damage that is observed. A rotating updraft tends to surge higher in the atmosphere and stay intact. The same winds cause the downdraft to stay distinctly separate from the updraft. The coupling of the updraft and downdraft is an orderly but extreme machine that is capable of continually ingesting warm, moist air and

Vertical Development — Clouds

increasing in intensity during its relatively long life cycle. Supercells can last for hours if the meteorological conditions are favourable for their formation.

A wall cloud is formed when the updraft draws in air from the area where precipitation has already fallen. These moistened air parcels are lifted upward into the updraft and condense to form cloud at a lower level than the lifted condensation level of the air mass—the base of the parent supercell cumulonimbus. These cloudy air parcels are drawn into the updraft and combine to form the mass of cloud below the updraft of the supercell—the wall cloud. Wall clouds are common and are not exclusive to supercells; only a few actually produce tornados. A tornado, should it develop, will typically originate from the wall cloud. The tornado is typically made visible by a condensation funnel, but it is the winds that cause the damage, not the funnel cloud. The damaging winds and not the cloud comprise the tornado. Damage often occurs on the ground even though the condensation funnel originating from the wall cloud may remain elevated well above the surface.

Mammatus clouds are bulbous or pillow-like cloud formations extending from beneath the anvil of a thunderstorm. These clouds form as cold air in the anvil region of a storm sinks into warmer and drier air beneath it. Mammatus are not exclusive to supercells and can be associated with any cumulonimbus. They can be very unusual and beautiful cloud formations.

Chapter 3: Precipitation

Water vapour is present even in the driest of air masses, but it is the conversion of vapour into liquid or solid forms that creates much of the weather we experience. A simple analogy for how water vapour results in precipitation can be made with a sugar-saturated cup of hot coffee. When the solution cools, the liquid cannot continue to retain all of the sugar in its dissolved state, and some precipitates out as solid crystals.

Water vapour behaves like any other gas in the atmosphere, but atmospheric water has the ability to convert between solid, liquid and gaseous phases. The atmosphere cannot hold an unlimited amount of water vapour, and the amount of vapour that can be held is entirely dependent on the temperature of the mixture. In the atmosphere, if the air at a given temperature is saturated with water vapour, cooling results in condensation into water droplets and sometimes sublimation directly into ice crystals.

Air does not contain a large amount of water vapour. One cubic metre of air at sea level, with a temperature of 35° C, weighs about 1 kilogram. If that cube is saturated with water vapour, the water has a weight of only about 37 grams. We can measure atmospheric water vapour content in a number of different ways. Vapour pressure is the actual pressure exerted by water vapour, and it is always a small fraction of the air pressure. At a given temperature, there is a maximum vapour pressure that can be reached before a phase change occurs with the excess. This process is known as the saturation vapour pressure.

Precipitation

Condensation in the atmosphere does not happen readily because the air molecules are always in motion. To stick together and form a liquid, the molecules require the presence of condensation nuclei. These nuclei are small, ranging from 0.1 to 10 microns (thousandths of a millimetre) in diameter. The lower atmosphere contains between ten thousand and one million condensation nuclei per cubic centimetre. In the upper atmosphere, there are only hundreds per cubic centimetre. For ice crystals to form, freezing nuclei are required. However, freezing nuclei are far less numerous than condensation nuclei, and there may be only one or two per cubic centimetre. Most freezing nuclei only become active at temperatures below -4° C, and some freezing nuclei do not become active until the temperatures are -10° C or colder.

Condensation forms cloud droplets that are typically 10 to 20 microns in diameter. When raindrop sizes are measured, they are typically found to be about 2000 microns in diameter (or 100 times larger than the cloud droplets). So, a typical raindrop is made up of more than one million cloud droplets. A major question for atmospheric scientists used to be what process occurs in the atmosphere to form raindrops from cloud droplets.

Official Definitions of Precipitation

Drizzle
Drops with diameter less than 0.5 mm, falling close together. They appear to float in air currents, but unlike fog, do fall to the ground.

Light Drizzle
Visibility more than 1 km.

Moderate Drizzle
Visibility from 0.5 to 1 km.

Heavy Drizzle
Visibility less than 0.5 km.

Rain
Drops larger than 0.55 mm or smaller drops that are widely separated.

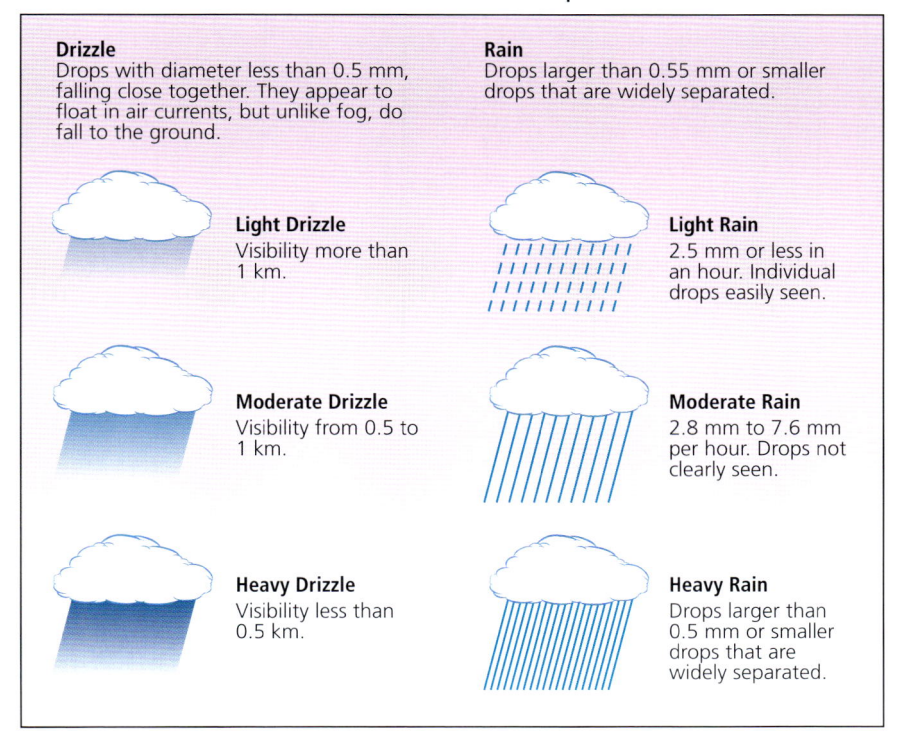

Light Rain
2.5 mm or less in an hour. Individual drops easily seen.

Moderate Rain
2.8 mm to 7.6 mm per hour. Drops not clearly seen.

Heavy Rain
Drops larger than 0.5 mm or smaller drops that are widely separated.

Fig. 3-1

Precipitation

The Raindrop Formation Process

Precipitation forms in the presence of water vapour if cooling of air also happens. First, the cooling. For most cloud formation, this process usually occurs when a parcel of air is lifted. This lifting can happen when an air parcel is heated near ground level and becomes less dense than neighbouring air parcels. Another lifting process occurs when air moves over elevated terrain. Also, if an air mass moves horizontally against a cooler air mass, the air masses do not readily mix, and the warmer air rides over the colder air along the boundary between the air masses, which is called a front. This lifting process is known as frontal lift. The rate of cooling also depends on whether or not the air parcel is saturated with moisture. If the air parcel is not saturated, it cools at a rate of about 10° C per kilometre. A water-saturated parcel cools slower, at a rate of 5° to 7° C per kilometre.

Two processes are thought to lead to the formation of raindrops. In the 1930s, scientists Alfred Wegener, Tor Bergeron and Walter Findeisen developed a theory about a process that is dependent on the presence of ice crystals. Both ice crystals and water droplets can form at temperatures below freezing, but when a mixture of water droplets and ice crystals is present, the ice grows more rapidly than the water droplets. In fact, water droplets do not spontaneously freeze until temperatures are below -30° C. So, at cold temperatures, ice crystals spontaneously form and grow at the expense of water droplets. Gravity causes these ice crystals to fall. The cloud droplets also fall but at a much slower speed. The fall speed of droplets and ice crystals is dependent on their so-called terminal velocity. As they fall, the crystals continue to grow in the water-saturated air, and they collide with cloud droplets that have a slower terminal velocity. These droplets immediately freeze when they strike the ice crystals, and splinters of ice may be ejected, which may also grow into falling crystals. The ice crystals fall and, assuming that their fall speed is greater than the upward

Growth of Ice Crystals

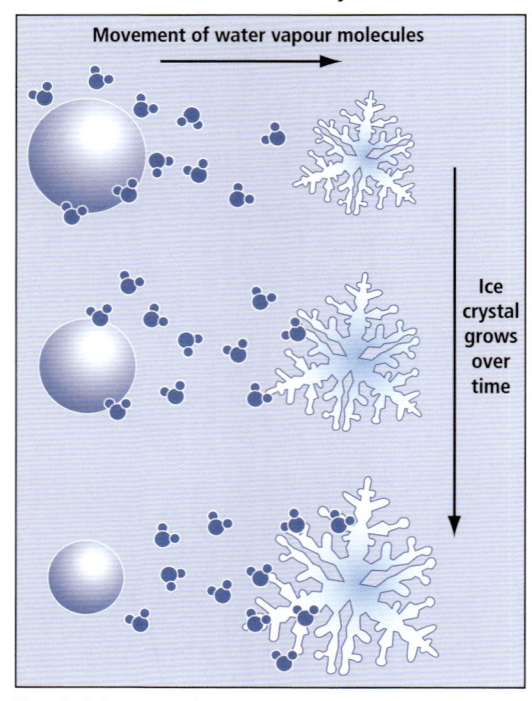

Fig. 3-2 Ice crystals grow more rapidly than water droplets when water vapour is abundant.

Precipitation

speed of the surrounding air, they move toward the ground. When they pass into air that is above freezing, they melt and form raindrops. This process is the "ice crystal theory," also known as the "Bergeron process," and seems to explain the formation of snow and most rain.

The ice crystal theory, however, does not explain the formation of rain in clouds that do not contain below freezing temperatures. In the tropics, and even occasionally at mid-latitude locations such as Ontario, precipitation falls from warm clouds. A second theory known as "collision theory" explains this rain formation process. In this theory, the terminal velocity of different sized cloud droplets has to be considered. The larger cloud drops fall faster and overtake the smaller droplets, and the raindrops grow by accretion. With the warm cloud accretion process, the rate of raindrop formation and the size of the resulting drops are directly proportional to the depth of the cloud. In Ontario, the precipitation formed in warm clouds usually consists of small drops that are known as drizzle.

Of course, the two precipitation-forming processes are usually underway at the same time in the atmosphere. In clouds that have significant depth, such as nimbostratus or cumulonimbus, both processes are happening within the cloud. Snow and ice crystals that form in a higher layer with temperatures below freezing, such as cirrostratus, may fall into a lower layer of cloud that is, in part, above freezing. This snow can stimulate production of raindrops in the lower cloud, which would otherwise not be thick enough to form raindrops.

Raindrop Growth by Accretion

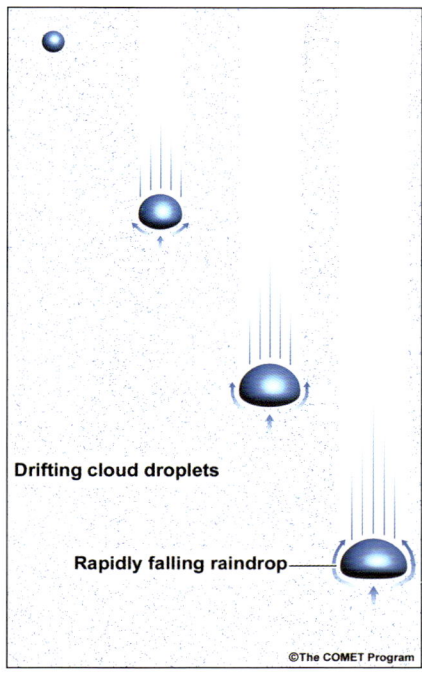

Fig. 3-3

Types of Hydrometeors

Rain is any liquid precipitation that reaches the ground. Raindrops can range in diameter from 0.5 to 7 millimetres. They are spherical in shape if they are less than about 2 millimetres in diameter, but larger raindrops falling through the atmosphere are flattened along the bottom because of the upward force of air. These larger drops resemble the top half of hamburger buns. This force also causes drops larger than 7 millimetres across to break into smaller, more spherical drops. Terminal velocity is achieved when the acceleration caused by gravity is balanced by the drag of the surrounding air. Large raindrops have a terminal velocity of almost 10 metres per second.

Precipitation

Snow Shapes: Basic Structure

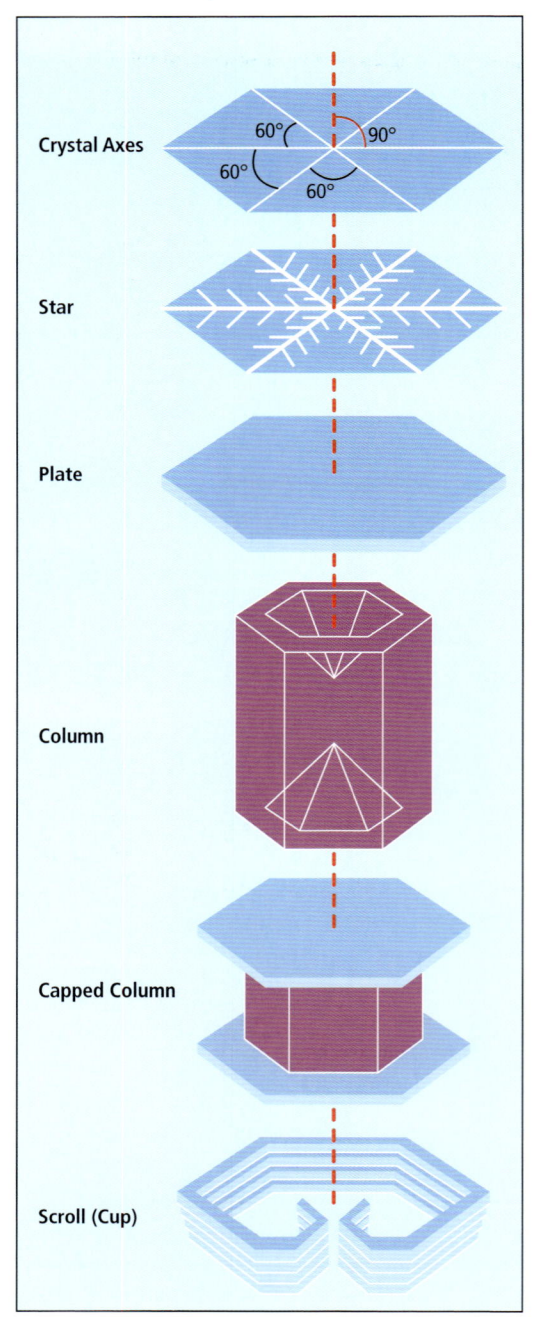

Crystal Axes — 60°, 60°, 60°, 90°

Star

Plate

Column

Capped Column

Scroll (Cup)

Drizzle consists of small drops that are less than 0.5 millimetres in diameter. These drops fall out of relatively thin cloud decks and are typically associated with fog. Drizzle drops have slow fall rates of about 2 metres per second, and even the slightest breeze causes drizzle to appear to swirl horizontally and not reach the ground. If temperatures are just below freezing at ground level, the supercooled drizzle drops freeze on contact. The accumulation of ice from freezing drizzle can be significant and can result in treacherous travel conditions on our highways. In Ontario, freezing drizzle conditions are most frequent during late winter and spring.

Snow crystals come in a variety of shapes and sizes. Snow in the atmosphere forms shapes that are dependent on the nature of the water molecule. The water molecule, which consists of one oxygen and two hydrogen atoms, has a polarity based on the positive and negative charge distribution. When water molecules link together in an ice lattice, opposite electrical polarities of the atoms attract, forming a solid ice crystal that has a hexagonal symmetry. The hexagonal building block grows in one of many characteristic

Fig. 3-4 The different shapes depend on the crystal structure of ice and the relative growth rates on each crystal axis.

Precipitation

fundamental shapes, which makes it possible to categorize snow crystals into several different classes. Figure 3-4 shows this crystal structure and the shapes of the various snow crystals and flakes that can result.

Snow or ice crystals that form at relatively warm temperatures between -2° and -10° C tend to be needles or columns, which can pack tightly together so that the snow, when melted into its water equivalent, yields a great deal of water. Liquid-to-snow ratios can get as low as 1 to 5. This is very wet, heavy snow, and it poses a severe health risk for those who try to shovel it. This type of small ice crystal snowflake is common with the large synoptic scale storms that have an ample supply of moisture and gentle lifting energy that slowly converts the moisture into snow. The precipitation growth area does not get high into the colder levels of the atmosphere. These large storms have equally large precipitation areas so that a long duration of the snow at a particular location can really cause even the most densely packed crystals to pile up.

Fig. 3-5 Temperature is the primary factor that controls shape, but available water vapour also has an influence.

Snow Shapes: Temperature Determines Shape

0°C to -2°C
Thin plates

-2°C to -6°C
Needles

-6°C to -10°C
Hollow columns

-10°C to -12°C
Sector plates

-12°C to -22°C
Dendrites

-16°C to -22°C
Sector plates

Below -22°C
Hollow columns

59

Precipitation

As the ice crystals form at ever-colder temperatures between -10° and -22° C, a dramatic change occurs. The ice crystals change to plates and dendrites. These larger, classic snowflakes do not pack together well, and the snowfall depth increases rapidly. When melted into their water equivalency, dendritic snowflakes yield much less water. Liquid-to-snow ratios climb to 1 to 15, or so. At the colder temperatures for snowflake creation, liquid-to-snow ratios can even exceed 1 to 20. This light, dry, fluffy snow accumulates quickly and is easily blown around. This type of large dendritic flake snow is typical of unstable air masses with convective clouds in which the precipitation growth area is higher and thus colder than is common in the larger synoptic scale storms. These convective clouds are also typically smaller in spatial and temporal extent. They tend to sweep past a given location quickly. As a result, though dendritic flakes can pile up quickly, neither the duration nor the spatial extent of the convective cloud is sufficient to result in severe accumulations. The exception to this rule is the snowsquall, a winter storm phenomenon explained in Chapter 8.

At temperatures colder than -22° C, the ice crystals return to hollow columns. These hollow columns do not accumulate, and at these cold temperatures, there is little moisture anyway.

It is probably true that no two snowflakes are alike. No two snowflakes have the same temperature, moisture or collision history on their long journey through the atmosphere. As snow falls toward the ground, it can take on a variety of shapes. Perfectly symmetrical snow

How Graupel and Hail Form

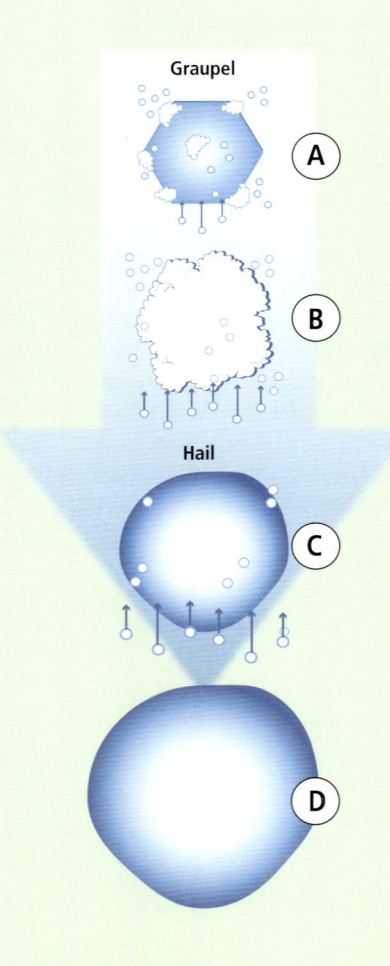

Fig. 3-6
A - supercooled cloud drops freeze to the crystal
B - cloud and raindrops freeze to form graupel
C - hail forms as more liquid raindrops freeze
D - hailstone falls out of the cloud when the updraft can no longer support its weight

Precipitation

crystals are rare. Usually the air is at least somewhat turbulent, and the branches of stellar flakes can break off. Snow crystals may break as they strike each other, or they may stick together. Cloud droplets also strike the snowflakes and may freeze instantly, forming what is known as a rime coating. This causes the original crystal or flake to take on a lumpy, irregular shape.

When ice crystals are subjected to a lengthy riming process as they fall through deep cloud, they form **graupel** or **snow pellets**. These heavily rimed ice particles are sometimes called soft hail, and they are usually no bigger than 5 millimetres across. They may take on a conical shape. Snow pellets can be seen at ground level when temperatures are at or slightly below freezing. When they hit

Large hail samples compared to a dime, a nickel and a quarter.

a hard surface, they may bounce but often shatter. Formation of graupel is important in the cloud electrification process. Graupel formation in deep convective clouds is the main mechanism for the formation of charged regions that lead to the generation of lightning.

Freezing rain (accretion)

Precipitation

Formation of Ice Pellets and Freezing Rain

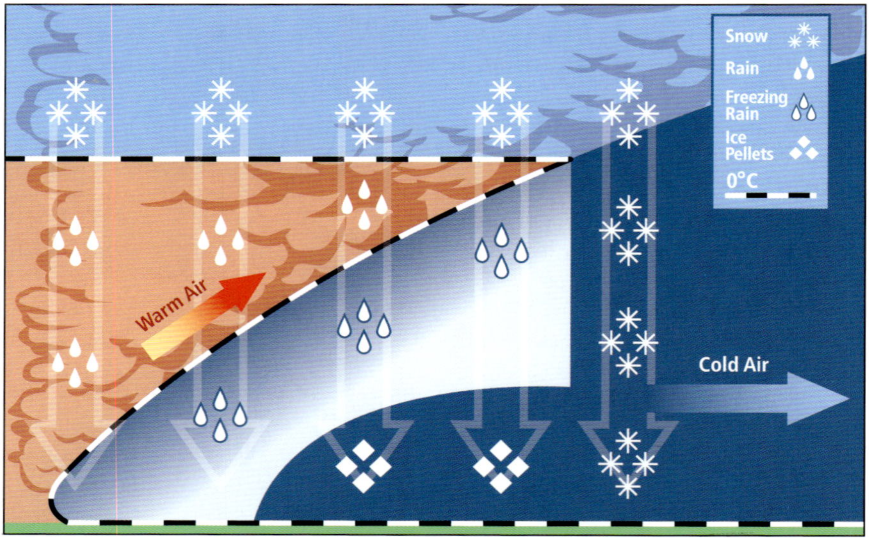

Fig. 3-7 Freezing rain, ice pellets or snow depend on the thickness of the sub-zero air mass.

Snow grains are small, opaque particles of ice that consist of bundles of rime or ice crystals held together by frozen cloud droplets. They fall at a light, steady rate from layers of relatively thin cloud and are considered to be the solid equivalent of drizzle. They usually have diameters of less than 1 millimetre.

Ice pellet is the name commonly given to precipitation consisting of refrozen raindrops or refrozen, partially melted snowflakes. The particles of ice are usually transparent, or nearly so. Ice pellets as observed at ground level are formed when raindrops refreeze after falling through a layer of cold air near the ground surface. They are 1 to 5 millimetres in diameter.

When there is a significant layer of warm air aloft and a thin layer of cold, sub-zero air at the ground surface, **freezing rain** results. In this case, the surface-based cold air is not deep enough to refreeze the falling raindrops, and the raindrops become supercooled drops of liquid that freeze almost instantly on contact with the surface. Even a brief period of freezing rain can disrupt transportation. As a result, freezing rain is arguably the most significant hydrometeor seen in Ontario. Airports, in particular, are keenly attuned to the possibility of freezing rain because a coating on runways can shut down airports for lengthy periods, especially if a prolonged cold spell follows a freezing rain event.

Hail is a very destructive hydrometeor. In their simplest configuration, hailstones are more or less spherical and are made up of concentric layers of clear and opaque ice. They form within actively growing convective shower clouds. Usually the electrical charge separation process is also underway in such clouds, so most hail occurs in association with lightning.

Precipitation

In active and developing convective clouds, strong upward and downward air movement can exist side by side. An individual hailstone begins its life as graupel that is supported by the strong updraft in the convective cloud. In strong updrafts, the graupel particle may rise a significant distance—perhaps several kilometres above its origin level. As it moves upward, it passes through regions that have high moisture content and a dense concentration of supercooled cloud droplets. With the high rate of collisions between the droplets and graupel, freezing is slowed for part of the accreted liquid, and the graupel becomes impregnated with water that freezes relatively slowly. The resulting pellet is more or less transparent. The growth of a hailstone is illustrated in Figure 3-6.

The hailstone is buoyed by the strong updraft and can grow to a significant size as the supercooled water-saturated air flows past. In upper regions of the cloud, the updraft speed may reduce or the hailstone may move into an area with reduced uplift. When the stone becomes too heavy to be supported by the updraft, it begins a downward trajectory toward the earth's surface, all the while sweeping through supercooled droplets and continuing to grow. Sometimes, the hailstone falls back into the stronger updraft and again moves upward. Figure 3-8 illustrates how hail growth occurs in a mature thunderstorm cell that is typical of a cold front. Updraft regions can be one kilometre or more across, and speeds in the most intense regions have been measured reaching 30 metres per second

Structure of a Hailstorm
(Cold Frontal or Squall Line Thunderstorm)

Fig. 3-8
A - hailstones grow in weaker updraft and fall out as medium-sized hail
B - large hail rises higher in cloud in strong updraft region

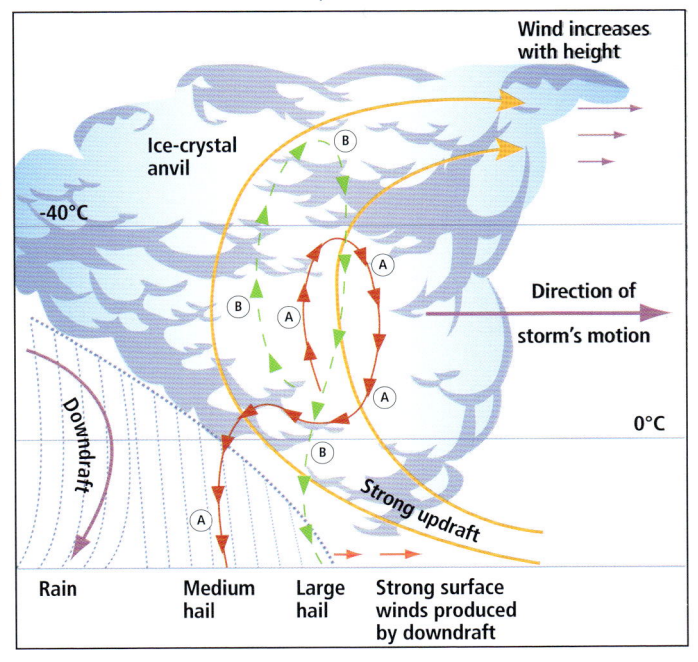

63

Precipitation

Hailstone Cross Section

Different layers in a hailstone reveal different temperature and water vapour conditions at each growth stage.

(108 km/h) or more. The largest hailstones grow in these environments. Large hailstones often have a lumpy structure, possibly because the hailstones are spinning and tumbling in the updraft, and water streams off the stone, forming icicle-like projections. Many hailstones have an onion-like structure, with many layers that formed as the stone moved between different updraft regions with different temperatures and liquid water content. This structure is revealed when cross-sections of hailstones are examined, as shown in the photo above. Extremely large hailstones often consist of smaller individual stones that have collided and become frozen together within the storm cloud.

Virga

Virga is defined as streaks of precipitation that fall from the base of a cloud but disappear before reaching the earth's surface. This phenomenon deserves special mention because it occurs with almost every synoptic scale system crossing Ontario. Weather systems laden with moisture that approach from the southwest can produce hours of virga before the first precipitation actually makes it to the ground.

Precipitation falling from high above a warm frontal surface falls into the drier, cooler air below. Typically at mid-latitudes, this precipitation forms as snow in the higher, colder reaches of the atmosphere.

The snow falls at about 1 metre per second. This falling snow might sublimate directly into water vapour or could melt into faster falling, small raindrops. Both effects result in a lower density of hydrometeors and the apparent disappearance of the precipitation to a ground-based observer. For virga, the small raindrops must evaporate before reaching the earth's surface. If they don't fully evaporate, the result is rain.

Precipitation Transitions

With a typical winter storm crossing Ontario, precipitation starts as light snow and then transitions to heavier snow. As the warm front approaches, the snow may change to ice pellets and possibly even freezing rain. With the continued approach of the warm air, the precipitation should change completely to rain. Within the warm sector of the winter storm, the rain tapers off to a drizzle or ends completely. As the warm sector passes by, the cold front ushers in freezing air and snow flurries that typically taper off as a ridge of high pressure replaces the winter storm.

Occasionally, the typical precipitation transition with a winter storm can be completely reversed. If the temperatures in the lower atmosphere are above freezing, snow falling into the warm layer will melt into rain before reaching the ground. The air that supplied the heat to cause this melting will be cooled, which often reduces the temperature of the atmosphere adjacent to the ground to below freezing. The falling rain transitions into significant freezing rain when the surface temperature drops below freezing. Continued cooling can increase the depth of the surface-based freezing layer and transition the freezing rain to ice pellets. Typically, falling snow aloft cools the entire air mass, and the temperatures at all levels drop to near or below freezing. There is no longer any melting of the snow aloft, and precipitation that started as snow in the upper atmosphere reaches the ground unchanged as snow. In this way, the transition of precipitation types can be completely opposite to the typical sequence of hydrometeors expected with a winter storm. This situation happens most frequently when surface temperatures are near freezing and extensive virga is observed with the approach of a winter storm.

"Sleet" is not a term used by Environment Canada. The term is used in the United States to describe a mixture of rain, snow and ice pellets. Environment Canada explicitly mentions all precipitation types that may occur at the same time.

Chapter 4: Atmospheric Electricity

Lightning

The atmosphere has both an electrical structure and a physical structure. In the lower portion of the atmosphere, lightning plays a key role as an electrostatic generator that helps recharge two concentric conductors—the surface of the earth and the electrosphere, which consists of highly conductive layers at altitudes above 50 to 60 kilometres. In the upper atmosphere, above 80 kilometres, the electrical phenomenon of most significance is the aurora borealis, which results when particles emitted by the sun interact with the atmospheric gases at those levels. There is not thought to be any interaction between the electrical processes in the lower and upper atmospheres.

The earth's surface has a negative electrical charge, and the lower atmosphere has a positive charge. This positive charge flows steadily to the earth's surface, in fair-weather conditions. In fact, if nothing else happened, the earth's charge would be neutralized in approximately 10 minutes, depending on the presence of conducting polluting gases. The thunderstorm acts as a generator that maintains the positive electrical charge in the atmosphere by means of the lightning discharge.

Atmospheric Electricity

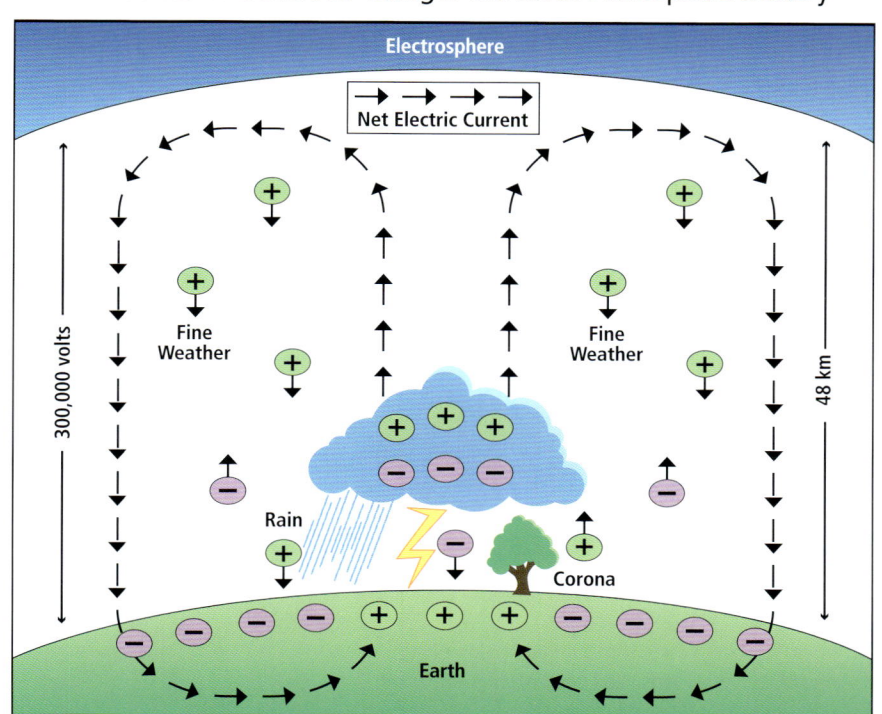

Thunderstorm Generator Charges the Earth-Atmosphere Battery

Fig. 4-1 Thunderstorms act as generators to keep the earth charged negatively and the atmosphere charged positively.

Toronto's CN Tower is hit by lightning 40 to 50 times per year. So much for "lightning never strikes in the same place twice."

The Cloud Electrification Process

Two mechanisms are thought to be responsible for electrifying clouds, turning them into thunderstorms—the convection mechanism and the graupel-ice mechanism. Although the latter process is the most widely favoured, the two processes may work in tandem, and there may even be an electrification process that is still unrecognized.

With lightning, the "leader" stroke is the one that reaches from the cloud to the ground, setting up a path for the "return" stroke, which travels from the ground up to the cloud. It is the return stroke that we actually see.

Atmospheric Electricity

Convective Mechanism of Cloud Electrification

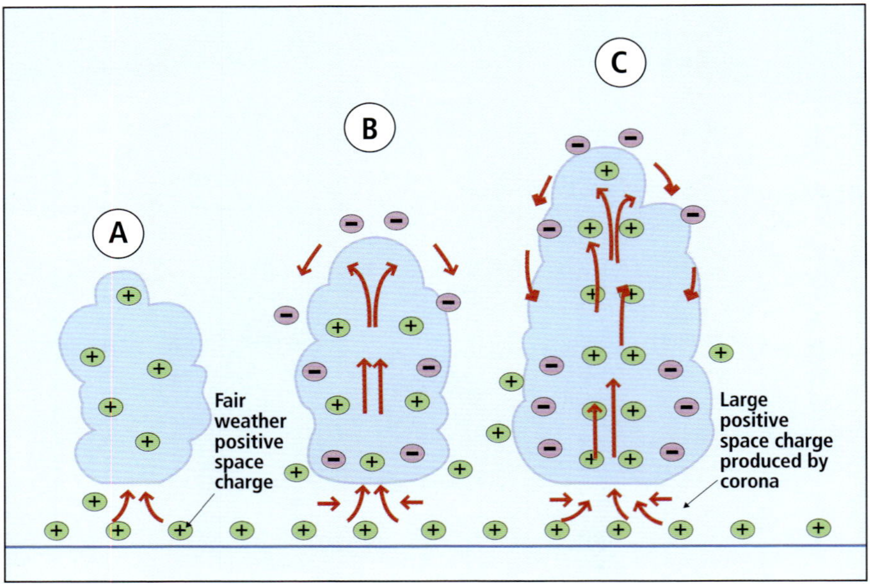

Fig. 4-2
A - positive charge swept into a developing cumulus
B - negative charge attracted toward edge of cloud
C - downdrafts on edge of cloud transfer negative charge to cloud base

The Convective Process

Convective clouds go through a growth and intensification process that was described earlier. Convective clouds consist of updrafts of air parcels that are buoyant relative to the adjacent air mass. These rising parcels contain moisture that condenses, forming the visible cumulus cloud. In the convection theory, shown in Figure 4-2, external sources provide the electrical charge. Updrafts at the cloud base carry positive, fair-weather charges toward the top of a growing cumulus cloud, and the charge is more concentrated than that in the surrounding air. Negative charges produced by the constant bombardment of cosmic rays are attracted to the positive cloud and, in turn, attach to the cloud particles at the outer portions of the cloud. These negatively charged cloud parcels move downward in the convection circulation at the edge of and outside the cloud, compensating for the convective updrafts inside the cloud. This negative charge moves toward the cloud base where it can discharge to earth, producing a positive corona, which produces a positive charge under the cloud and a positive feedback process. This process does not explain all the complex charge distributions in fully developed thunderstorms.

68

Atmospheric Electricity

The Graupel-Ice Process

The graupel-ice process cannot function unless a precipitation process is present in a growing cloud. The precipitation elements are graupel particles. As gravity draws graupel particles toward the earth, they collide with cloud droplets and ice crystals that are at the same levels within the cloud. When collisions occur, the graupel particles, as well as nearby ice crystals, acquire either a positive or a negative charge. The sign of the charge depends on the temperature of the air, the water content of the cloud, ice crystal and water droplet sizes, the relative speed of particles in the collisions and the chemical pollutants in the water.

The most important factor is the temperature at which the collisions occur. In the upper part of the cloud where the temperatures are colder, the falling graupel acquires a negative charge and the crystals acquire a positive one. At warmer temperatures, the graupel is positively charged and the adjacent ice crystals are negatively charged. The critical temperature is between -10° and -20° C.

This charge-separation process then results in what is known as the tripole distribution observed in thunderstorms (see Figure 4-4). In this tripole, a positive charge is located at the top of the cloud (known as the P-region), a large region of negative charge is in the lower half of the cloud (known as the N-region) and a smaller region of positive charge (called the p-region) is at the cloud base.

The Graupel-Ice Charge Separation Process

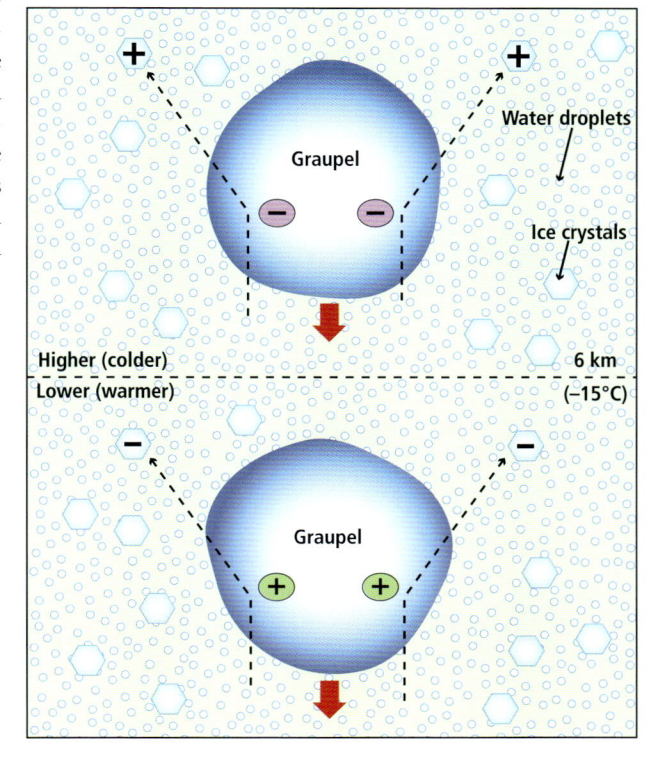

Fig. 4-3

Atmospheric Electricity

Charge Distribution in a Thunderstorm Cloud

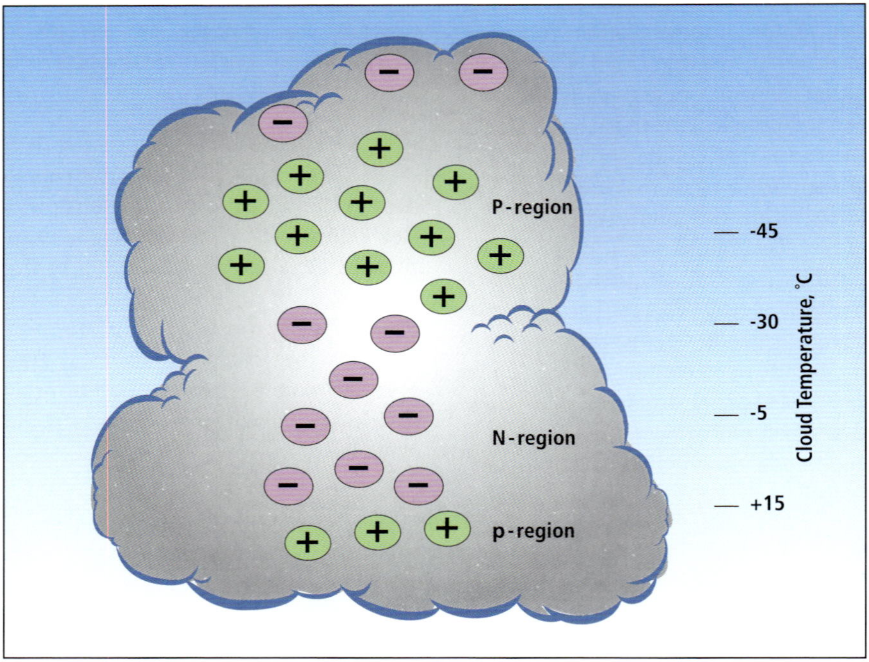

Fig. 4-4 The likely charge distribution in a fully charged thunderstorm

Generation of a Lightning Stroke

There are two primary types of lightning stroke: cloud-to-ground strokes (from the charged portion of the cloud to the earth) and intracloud strokes (between charged portions of a cloud). Occasionally, strokes pass between the charged portions of two clouds (called cloud-to-cloud lightning), and even more rare is a strike from cloud to air. During thunderstorms in North America, there are about five times as many intracloud discharges as cloud-to-ground discharges. Most of the pictures of lightning strikes are of the cloud-to-ground variety because the intracloud flashes are usually obscured by the cloud itself.

Although the charged cloud typically has a triple configuration, the N-region is of greatest importance in cloud-to-ground discharges. As the cloud moves over the land surface, this negative charge region induces the build up of a

Lightning can heat the air it passes through up to 30,000° C, five times the temperature of the surface of the sun.

70

Atmospheric Electricity

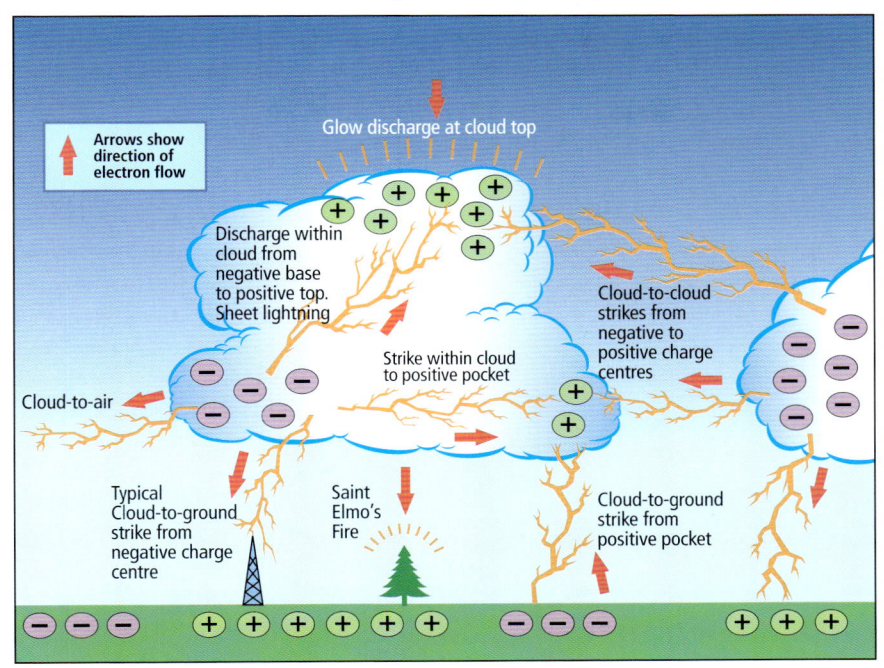

Coronal Discharges and Lightning Forms

Fig. 4-5 Saint Elmo's Fire is a coronal discharge that occurs when a thunderstorm is present. Many different types of lightning occur in thunderstorms.

positive charge at the earth's surface. This positive charge area stays under the cloud as the cloud moves along.

The lightning stroke mechanism is complex, but most scientists believe the process starts when the negative and highly mobile electrons in the N-region discharge to the small p-region immediately below. The electrons overrun the p-region and continue their trip toward the ground in a rapid series of bursts along a channel known as the stepped leader. The lower end of the stepped leader moves toward the ground at about 120 kilometres per second. As the tip of the stepped leader approaches the ground, a strong positive charge is induced, especially on objects projecting above the ground surface. Upward-moving positive discharges, called streamers, are initiated from these points, and soon the downward stepped leader and the upward positive streamer connect. Electrons at the bottom of the stepped leader rapidly discharge to the ground, and the lower portion of the channel becomes luminous. As electrons are pulled from higher and higher up the channel, the region of the strongest surge of downward rushing electrons also moves upward, as does the associated luminous region.

Atmospheric Electricity

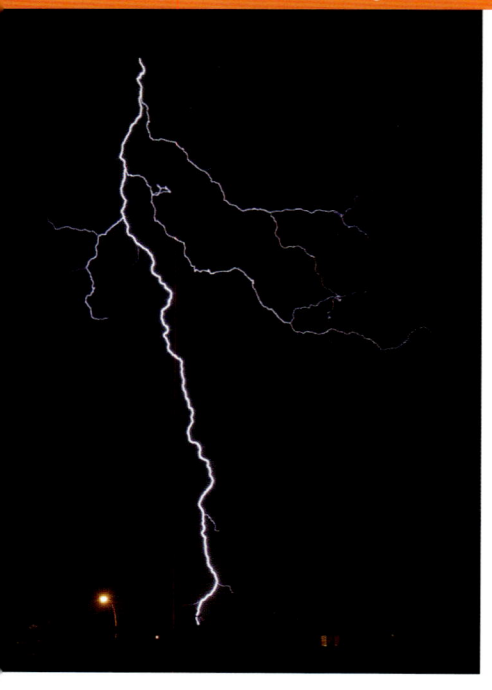

A cloud-to-ground lightning stroke

The upward movement of the luminous region moves at a speed of about 100,000 kilometres per second, or around one-third of the speed of light, and takes about 100 millionths of a second. This process is known as a single stroke discharge. The human eye is not capable of resolving the rapid speed of the luminous channel, and the flash appears to be one continuous stroke. Typically, however, several strokes occur within the same channel, progressively discharging higher and higher portions of the N-region. These consecutive strokes are typically separated by gaps of only 40 to 50 thousandths of a second. This rapid sequence of flashes gives lightning its flickering appearance.

The lightning discharge process is, of course, what is responsible for thunder.

The sound of thunder is a result of lightning rapidly heating the air it passes through, causing the air to expand explosively. The rapid flow of electrons in the lightning channel causes a rapid heating of the channel to about 30,000° C. This heating causes an explosive expansion of the channel. This expansion compresses the surrounding air, producing a shockwave that propagates rapidly in all directions. After only a short distance, the shockwave converts to a sound wave that propagates at the speed of sound, which near sea level is about 330 metres per second. For an observer, the light from a flash appears instantaneous. However, the sound moves much slower. For a cloud-to-ground discharge, we hear the sound from the closest portion of the channel first and from the most distant portion last. Therefore, depending on the orientation of the lightning discharge channel, the thunder duration may be spread out over several seconds. If the orientation of the discharge channel is perpendicular to your line of sight, the thunder sounds like a crack because the sound waves reach your ear at about the same time. Thunder is more of a rumble if the orientation of the discharge channel is along your line of sight.

Anatomy of a Lightning Stroke

Fig. 4-6
A - electrons accumulate in the base of the cloud
B - electrons move downward along stepped leader
C - streamer moves upward from an object on the earth's surface
D - upward moving positive charge creates a luminous discharge channel

Atmospheric Electricity

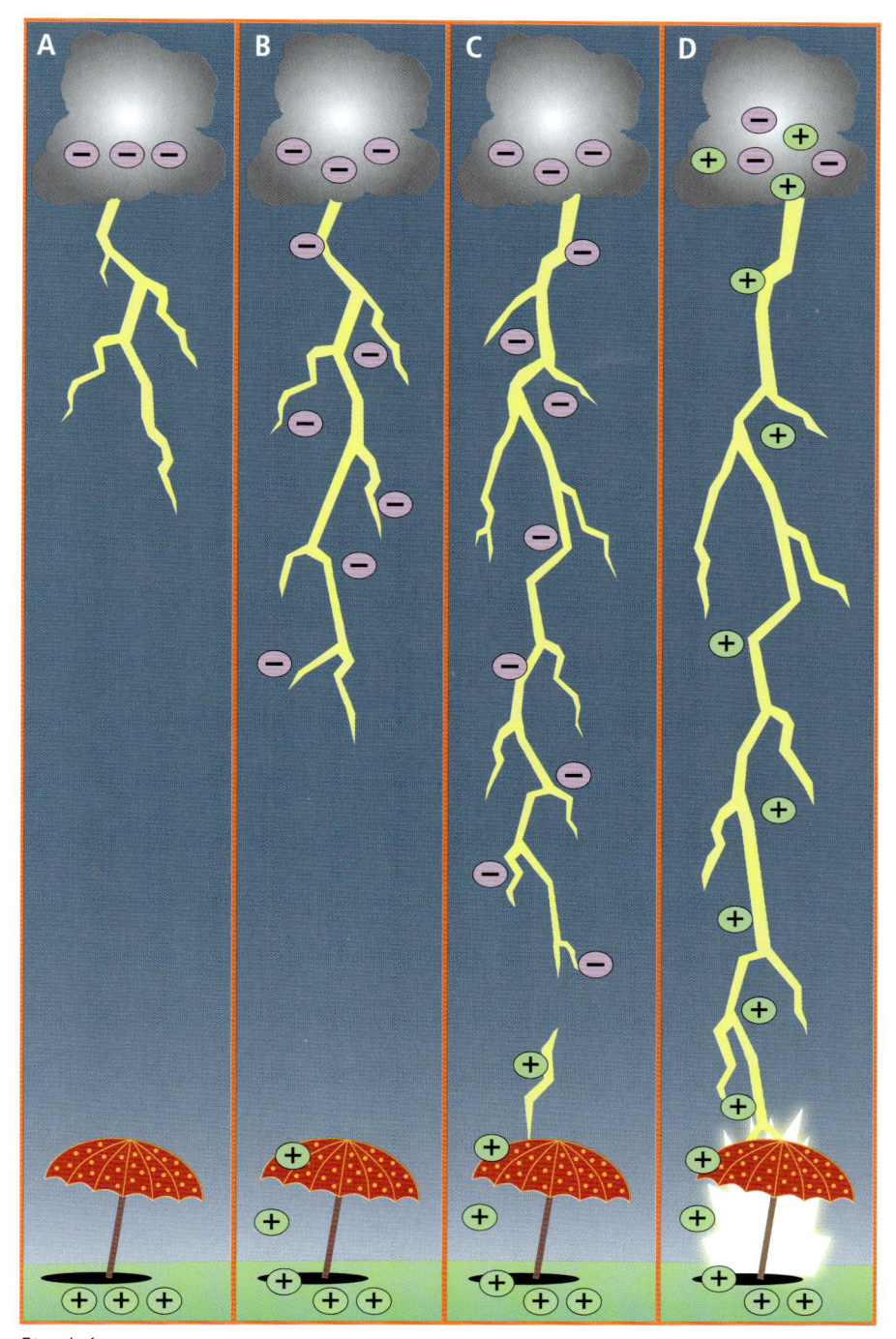

Fig. 4-6

Atmospheric Electricity

The pitch of the thunder depends on the intensity of the explosive heating and the altitude at which the thunder shockwave is generated. To add complexity, the atmosphere itself filters out higher pitched sound at a rate that is primarily dependent on the distance the sound has to travel through the air. Nearby thunder has a relatively high pitch, whereas more distant thunder has had a lot of the higher pitches suppressed and is a low rumble. In a nearby lightning discharge, we occasionally hear what sounds like hissing or cloth tearing. This hissing sound is believed to be a combination of the sounds made by the downward-moving stepped leader and the upward-moving streamers.

The sound of thunder is a result of lightning rapidly heating the air it passes through, causing the air to expand explosively.

The lightning process described has been concerned with discharges from the N-region in thunderstorms, but occasionally when the P-region is closest to ground level, it initiates a discharge. This can happen in a number of ways: if the cloud is strongly sheared such that the lower N-region is dispersed; if the cloud is relatively shallow and only develops a small N-region; or if the cloud is dissipating and the residual upper portion of the cloud remains intact. Thunderstorms that form in winter also have a higher frequency of positive lightning discharges because, in cold weather, the charge formation process favours establishment of a more extensive P-region. Lastly, a positive discharge can occur from a thunderstorm passing near a tall tower or elevated terrain.

Positive discharges typically consist of a single stroke, rather than the multiple strokes that are normally experienced. The positive strokes also carry a large amount of current to the ground. The magnitude of the total charge exchanges can be up to 10 times larger than those in a typical negative discharge. With these large current flows, positive strikes also have a greater tendency to cause combustion and may be a primary cause of forest fires, especially because they often take place well away from a precipitating region of the storm cloud.

Although most lightning is of the cloud-to-ground or intracloud variety, upward discharge has also been observed from the top of the thunderstorm. There have been reports, mainly by high-flying U-2 aircraft, of upward propagating lightning strokes that have the same appearance as commonly observed lightning. These discharges appear to be a few kilometres in length.

Other, more diffuse discharges have also been observed. Known as red sprites and blue jets, they extend 10 to 30 kilometres above large thunderstorms. They appear to occur in association with significant cloud-to-ground strikes. Although rare, their presence had been proposed as early as the 1920s. Sprites and jets are thought to form when a pulse of energy moves upward at the same time as a downward lightning discharge. This energy pulse, in turn,

Atmospheric Electricity

Multiple strokes in a time-lapse photograph of a thunderstorm

> Thunder is good, thunder is impressive; but it is lightning that does the work.
>
> —Mark Twain

causes electrons to break free of gas molecules in the region above the thunderstorm. A cascade of energized electrons moves upward and causes the oxygen and nitrogen molecules to emit light. A ring of green light may also appear to flash at altitudes of 80 to 100 kilometres above ground level in association with the red sprite. These phenomena are difficult to observe because clouds are usually blocking the view of a storm's top, and the actual discharge is faint.

Atmospheric Electricity

Ball Lightning

Small, luminous spheres sometimes observed during thunderstorms are called ball lightning. These balls are reported to be about the size of an orange or grapefruit, though some have been reported to be as large as 30 to 40 centimetres in diameter (larger than a basketball). The balls are usually located within a few metres of the ground. They move erratically, occasionally appearing to bounce along the surface. Lightning balls mostly appear yellow to red and seem to be rather innocuous. When they disappear after a short lifespan of seconds to one minute, they usually decay silently, though there have been reports of a small explosive sound. At present, there is no reliable theory for the formation and behaviour of ball lightning.

Ball lightning in a 19th-century woodcut

Areas of Lightning in Ontario

Around the world there are approximately 2000 thunderstorms in progress at any given time, and with the rate of cloud-to-ground lightning flashes in an average thunderstorm, there may be 30 to 100 flashes to ground every second. In Ontario, the thunderstorm season is generally from April to October, though thunderstorms sometimes occur outside this period. During summer, there is usually at least one thunderstorm in progress every day.

Lightning detection networks, established to assist in forest fire detection and management programs, have made it possible to map the location and frequency of cloud-to-ground and intracloud lightning flashes. The number of flashes is, of course, closely related to the intensity of an individual thunderstorm. The more severe the storm, the more lightning discharges occur.

Use the AM channel of your radio to detect a lightning stroke 30 km or more away. Increasing "crackle" on your AM band indicates increasing intensity of the updraft and the convection.

Atmospheric Electricity

Effects of Lightning Flashes

Forest Fires

Given the global rate of cloud-to-ground flashes, forests worldwide might be expected to receive about 50,000 flashes per day. Not all flashes initiate forest fires, but it is estimated that lightning is responsible for starting half of the wildfires in Ontario. During the 10-year period ending in 2006, there was an average of 1292 wildfires per year in Ontario, with 695 of these initiated by lightning. Studies of the characteristics of a typical cloud-to-ground flashes suggest that peak electrical currents are in the range of 10,000 to 20,000 amperes, with occasional peaks over 100,000 amperes. This current flows for a short period—usually less than one-thousandth of a second. Often there is a continuing current of 100 amperes or so that lasts for another one- or two-tenths of a second, and this type of flash, with a short continuing current, is believed to be responsible for starting combustion. Lightning appears to follow a path through a thin layer of living cells between the inner bark and the wood layer. With the passage of the large currents through the tree, there is explosive heating—a strip of bark is generally blown off, and the tree itself may be split or splintered. Depending on the dryness of the tree bark and ground cover, a fire may be ignited. In many cases, the fire does not start immediately, and combustible material below ground level may smoulder for a few days before surface combustion begins.

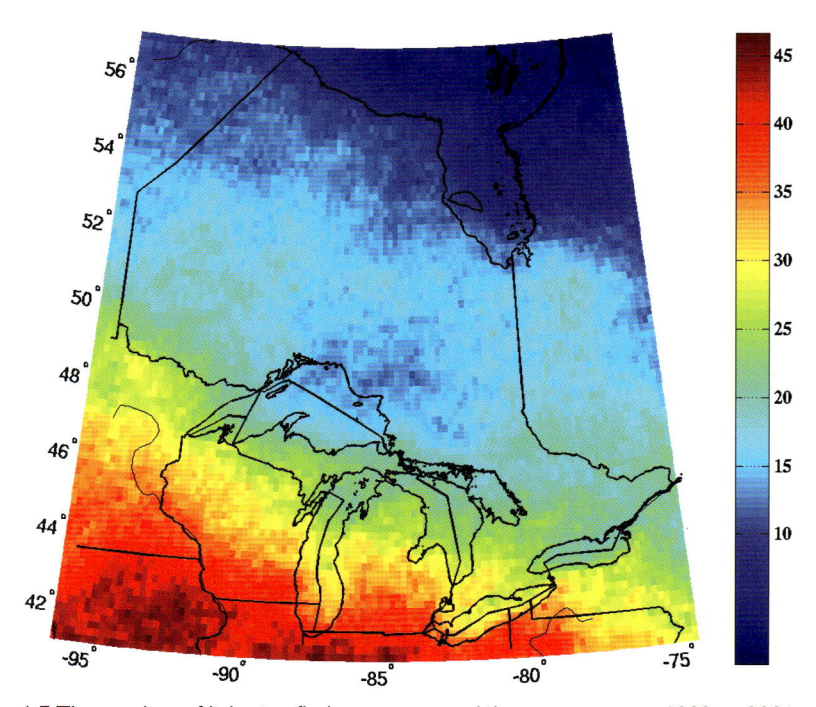

Fig. 4-7 The number of lightning flashes per square kilometre per year, 1999 to 2006

Atmospheric Electricity

Fulgurites can form when lightning strikes sandy surfaces (below).

Lightning is responsible for 5 to 8 percent of global nitrogen fertilizers. Lightning supplies the energy required to split the inert nitrogen molecules. The nitrogen atoms combine with oxygen in the air to form nitrogen oxides, which dissolve in rain to form nitrates. These nitrates are carried to the ground as fertilizer.

Fulgurites

A lighting flash can contain up to 100 million volts of electricity and carry 10 to 20 kiloamperes of current. When lightning hits a land surface that is made up of sand and certain kinds of rock, the heat can be high enough to melt the material along the lightning channel. This material solidifies, creating what is known as a fulgurite. Fulgurites take the form of long, hollow tubes that range from 1 to 5 centimetres in diameter. They have been traced in surface sand layers to depths exceeding 15 metres. Ancient fulgurites that date back more than 250 million years have been uncovered.

In Canada, lightning is most frequent in southwestern Ontario and is rare in the Arctic.

Lightning Protection

Given the destructive nature of the cloud-to-ground lightning flash, protecting property from lightning-initiated fire has a long history. Benjamin Franklin undertook experiments confirming that lightning had an electrical nature,

A lighting flash can contain up to 100 million volts of electricity.

78

Atmospheric Electricity

and he was probably the first to suggest, in 1750, that lightning rods should be placed on structures to guide lightning to the ground through a conducting wire. Today, lightning codes have been established that outline the characteristics of an adequate protection system for residences. As well, high voltage electrical transmission lines have lightning protection systems built in. Combustible fluids must also be protected to ensure that lightning or any related electrical arcing does not contact any air-vapour mixtures.

Passive Electrical Phenomena

A variety of other processes occur in the atmosphere, the results of which can be observed in the sky in Ontario. These processes occur at various layers of the atmosphere and have a number of causes. They, for the most part, result in the production of a weak, light glow and can be classified as having an optio-electrical basis.

Coronal Discharge

Coronal discharges occur any time there is a strong electrical field present in the atmosphere. Such strong electrical fields occur underneath active convective storms and cause ionization of the air molecules. When the gas is ionized, electrons in the molecule are elevated to a higher energy level. When the electrons return to their stable state, they emit light—a process known as fluorescence. This process is similar to the mechanism that causes neon lights to glow. In the atmosphere, the coronal discharge appears as a luminous bright blue or violet glow that can be accompanied by a hissing or buzzing sound.

Lightning Safety Tips

In the open
- stay at least 30 metres away from metal fences
- remove shoes that have metal cleats
- do not use metal objects such as bicycles, golf clubs or fishing poles (graphite rods)
- do not shelter under trees or canopies or in small sheds, picnic shelters and the like
- avoid open fields and high ground; if you are in an open field, crouch down and cover your ears; minimize your contact with the ground
- seek shelter in low-lying areas, such as valleys or depressions, that are not prone to flash floods

Indoors
- keep windows and doors shut, and do not approach them
- do not have a bath or shower or use tap water—electricity can be carried through the pipes
- unplug all electrical appliances
- do not use a phone that is connected to a land line

In a vehicle
- you are safe in an all-metal vehicle as long as you do not touch anything on the interior that is metal
- do not park near trees or under power lines
- if a power line should fall on or near your vehicle, do not get out of the vehicle
- convertibles are NOT safe—it is a vehicle's outer body that makes it safe, (it acts like a Faraday cage), not the rubber tires

Atmospheric Electricity

Electric fields are more concentrated in areas of high curvature, such as at the end of a lightning rod, ship mast, church spire, chimney or other pointed object. Aircraft flying through active electrical storms often develop coronal discharge streamers from antennas and propellers, and even from the entire fuselage and wing structure. The discharge can also appear on leaves and grass, from the tops of trees or thunderstorm clouds and even at the tips of cattle horns. Coronal discharge, also known as St. Elmo's fire, has been associated with ships at sea during stormy weather (Elmo is the patron saint of seafarers).

Aurora Borealis

All the electrical effects discussed to this point are associated with phenomena in the lower atmosphere. Ontarians, especially those north of the Great Lakes, are also frequent witnesses to the aurora borealis, which are observed in the night sky of winter. The aurora borealis occur in the very high atmosphere; they have no apparent influence on the weather we experience at the earth's surface, and they are sometimes put in a category known as space weather.

The earth has a strong magnetic field that is produced by the movement of the earth's molten core, and this magnetic field streams out into space. The sun is continually emitting a stream of positively and negatively charged particles known as plasma gas, which moves outward in what is often called a solar wind. Although the earth's magnetic field is greatly distorted by the pressure of the solar wind, the magnetic field serves the important function of keeping highly energized solar particles from reaching the surface of the earth. The upper reaches of the earth's magnetic field also contain electrically charged plasma gas. The theory of the formation of the aurora borealis holds that the flow of the sun's plasma past the earth's plasma causes electrons to become highly energized. These electrons strike the rarified atmospheric gases more than 80 kilometres above the earth's surface. These collisions impart energy to, or excite, the gas molecules. When the excited gas molecules return to their stable state, they emit light energy, which is the source of the aurora borealis. The colours we see are characteristic of the types of gases that are present in the atmosphere and depend on the energy of the particles that are stimulating the gas molecules. Atomic oxygen emits green and dark red light, and nitrogen emits blue and purple. The colours can be varied, and they shift depending on the motion of the magnetic field and the incident speed of the solar ions.

A recent study has shown that the more energetic aurora borealis are caused when the magnetic field on the night side of the earth is like a series of elastic bands stretched to their breaking point by the solar wind. When the magnetic field snaps back to normal, the stored "elastic" energy is flung back to earth, powering the brightest auroras.

The shape of the auroral pattern is directly related to the shape of the magnetic field at that level. The aurora borealis generally have a fairly sharp lower cutoff at their base height of 80 to 100 kilometres. The top of the pattern is generally indistinct, but it can extend several hundreds of kilometres above the base.

Atmospheric Electricity

Auroras are greatest at about 60° N, so the number of events and intensity of auroral displays improve the farther north the observer is located.

It has been suggested that the aurora borealis emit sound; however, the region of excitation is located at least 80 kilometres above the surface in an environment that is almost a vacuum. This makes it impossible for the sounds to come directly from the auroras. There are other theories to explain the origin of this sound. One theory suggests that the sound may actually come from electrical discharges on earth that are simultaneous with an aurora. Another theory proposes that the electromagnetic energy causes vibrations in objects such

Atmospheric Electricity

as pine needles or loose hair, thus causing the swishing or crackling noises described by spectators. Alternatively, the aurora borealis sound might result from a leaky optic nerve of the observer. The optic nerve, which is located adjacent to the part of the brain that is responsible for sound, interprets the leaked electrical impulses as sound. The apparent sound of the aurora borealis is said to vanish if the observer simply closes their eyes.

Air Glow

Air glow is a weak, blue-coloured emission by the atmosphere that forms as a result of several processes in the high atmosphere. During daylight hours, solar rays can cause molecules in the upper atmosphere to become ionized, and cosmic rays from all directions can excite atmospheric molecules. Chemical reactions between some gases present in the high atmosphere are also occurring. All of these processes result in the production of weak light energy. Although air glow is uniform across the sky, it is most readily observed toward the horizon because one is looking through a greater depth of the atmosphere, and therefore, the light is concentrated along those sight lines.

Chapter 5: Winds

Pressure differences in the atmosphere set the air in motion, and temperature differences generally induce pressure differences in the horizontal. In the atmosphere, there are other forces that come into play on air parcels as well.

The rotation of the earth causes all parcels to be deflected. As an air parcel moves, the Earth rotates under it, causing the parcel to follow a curved path on the earth. This apparent force is known as the Coriolis force (see pp. 17–18). In the northern hemisphere, air parcels appear to be forced to the right; south of the equator, they are deflected to the left. This apparent force is strongest at the poles and decreases to zero as you approach the equator. In the absence of other forces, the Coriolis force causes the winds to blow along lines of constant pressure. In the northern hemisphere, winds blow in a counter-clockwise direction around centres of low pressure and clockwise around centres of high pressure.

Any moving air parcel is also subjected to frictional forces, the most significant of which is the roughness of the underlying land surface. Frictional effects, which deflect winds by as much as 40 degrees toward lower pressure, are negligible above 400 metres over flat land or water surfaces and above 700 metres over rough terrain. Over water, the deflection is rarely more than 10 degrees. When looking at wind plots on maps of pressure at sea level, the drag of friction results in an apparent spiralling of winds toward the low pressure centres. The opposite is true for high pressure systems. Here friction

Winds

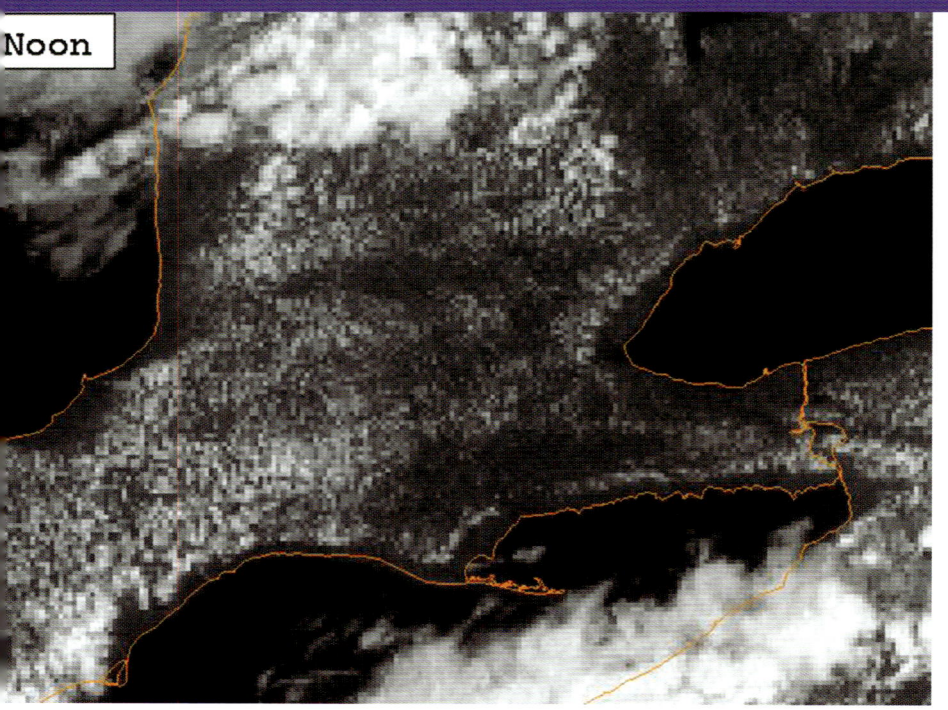

Noon

Satellite image of lake breeze cumulus

appears to cause an outward spiralling of air away from the centre.

Curvature in the flow results in another force that affects the movement of air. This force opposes the motion of air along a curved path. It is generally a slight effect that results in the movement of air outward and away from the centre of low or high pressure.

The aforementioned forces modify large-scale circulation and wind patterns. Geographical features such as mountains, valleys and large lakes have a multitude of local effects on wind patterns. In addition, small-scale differences in heating are responsible for several wind patterns.

Lake Breezes

Solar radiation is absorbed at different rates by land and water surfaces. Under sunny skies, air over land heats more quickly than air over an adjacent water surface. As daytime heating progresses, the difference between the air temperatures creates a corresponding pressure difference. The air over land has a lower pressure, and air over the water surface is pushed toward land by higher air pressure over the lake. The heated air rises, establishing a local circulation pattern known as a lake breeze circulation, as shown in Figure 5-1. The effect is most pronounced adjacent to large bodies of water such as the Great Lakes, but it is present around all lakes.

Winds

The circulation will progress during the day, strengthening and moving farther inland, and it can have a noticeable effect on the presence of clouds. Over land, the rising air parcels may have enough moisture to condense and form small cumulus clouds, whereas subsiding air in the return flow over the lake is generally cloud free. With sufficient air mass moisture, a line of cumulus will typically develop during the afternoon along the leading edge of the lake breeze. The flow pattern may also become visible if smoke or pollutants are present and become entrained in the air circulation. The lake air is typically less polluted, so visibility is better on the lake side of the lake breeze. The air off the water is also cooler, reducing temperatures in the onshore areas.

Lake Breeze Circulation

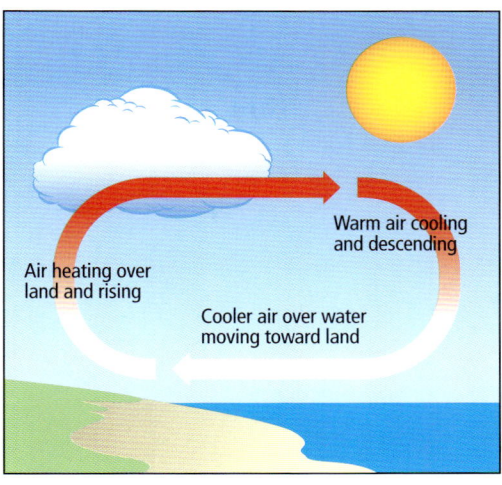

Warm air cooling and descending
Air heating over land and rising
Cooler air over water moving toward land

Fig. 5-1

Farther inland, temperatures will be warmer without the cool breezes coming off the lake. Lake breezes are most common in spring and summer when the water is at its coldest.

Land Breeze Circulation

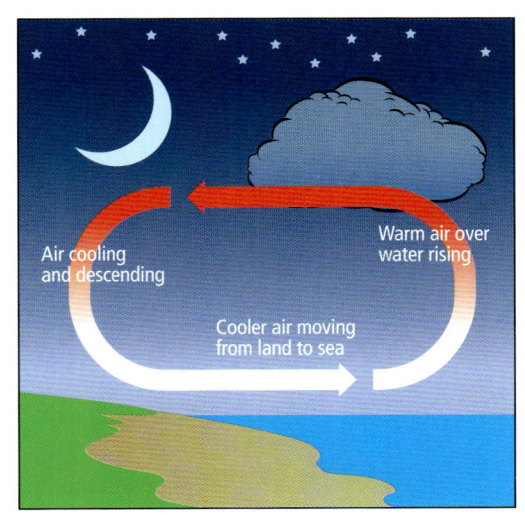

Warm air over water rising
Air cooling and descending
Cooler air moving from land to sea

Fig. 5-2

Land Breezes

The opposite effect occurs at night when solar heating has ceased. The land surface cools more quickly than the adjacent lake surface, and a point may be reached where surface pressures are higher over land than over the lake. Air is pushed offshore, establishing what is known as a land breeze circulation. This circulation is usually not as pronounced as the lake breeze. A line of cloud will often develop over the lake at night at the terminus of the land breeze. Land breezes form best with calm, overnight winds.

Winds

Terrain Effect Winds

Anabatic Wind

Katabatic Wind

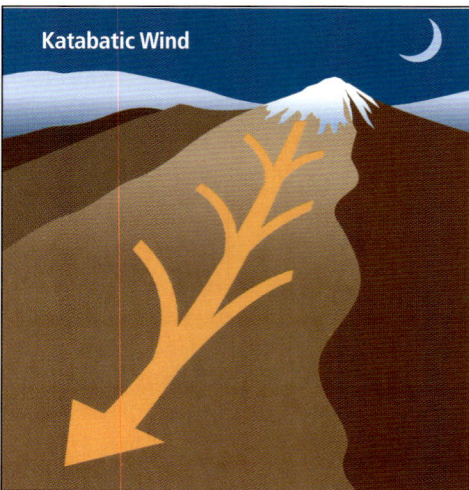

Fig. 5-3 Owing to heating and cooling, high terrain can substantially modify wind patterns.

> The pessimist complains about the wind, the optimist expects it to change; the realist adjusts the sails.
> —William Arthur Ward

Terrain Effects

High terrain can substantially modify wind patterns, often distorting what may otherwise be a uniform pattern based on the larger scale pressure pattern. Thermal circulations that are typically established in valleys and near mountains can also be observed in Ontario. The north to south Niagara Escarpment provides a subtle illustration of this phenomenon. The first cumulus of the day often go up on its eastern slope. During daylight hours, the sunlit side of a terrain feature may be strongly heated, and adjacent air will be significantly warmer than air farther out from the sloped terrain. The warmer air moves up the slope to higher elevations. This movement is called the anabatic wind pattern and can generate the escarpment cumulus. At night, a reverse circulation is established. In this case, the higher terrain feature radiates heat and cools the adjacent air more quickly than the air farther down the slope. The cool air begins to sink and flows down the sloped terrain, producing what is called a katabatic wind. This wind can be quite pronounced if the sloped terrain is snow covered and cools adjacent air quickly. Katabatic winds can occur day or night in winter. They are usually much more pronounced than anabatic winds, and the breezes that result can be quite cool.

The Jet Stream

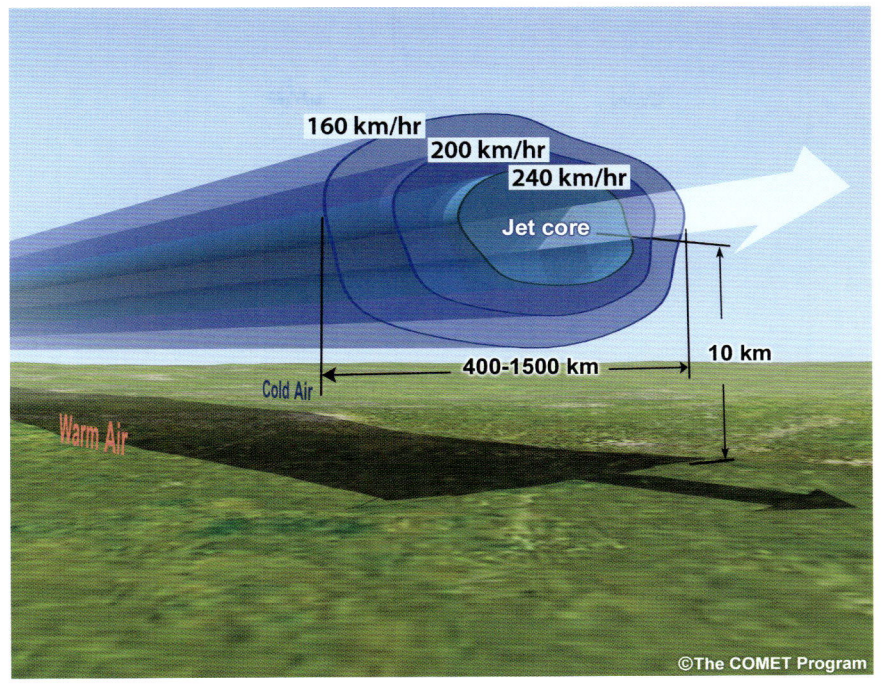

Fig. 5-4 Wind speed changes quickly above and below the jet stream core.

Jets

Large temperature differences in the atmosphere give rise to strong winds called jets. There are two common types of jets—the jet streams in the high atmosphere, and the low level nocturnal jets that appear in the lower atmosphere.

Jet Streams

The jet stream is located near the top of the troposphere, about 10 kilometres above the earth's surface. Jet streams at Ontario's latitude are usually associated with the frontal surface between air masses near the tropopause. At high altitudes, there may be several air masses, and therefore several jet streams, evident at any one time. Because the strength of the jet stream is a function of the temperature differences between air masses, the polar jet near the front that separates warm tropical air from colder, polar maritime air is usually the strongest. And because colder air is poleward of warmer air, the wind within the jet stream usually blows from west to east.

Jet streams are truly weather phenomena with long linear dimension. When weather maps covering the whole hemisphere are examined, the jet streams may be seen to encircle the entire hemisphere. Where vigorous weather systems are present in the lower atmosphere, the jet

Winds

Jet Stream Positions

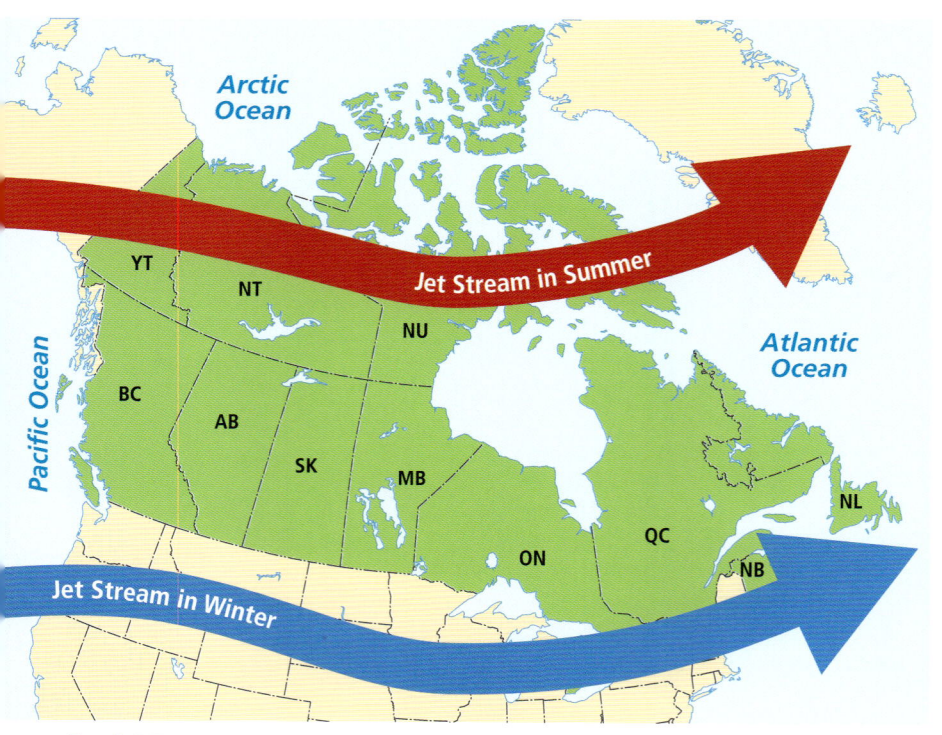

Fig. 5-5 The jet stream's average position moves north in summer.

stream appears to break into segments as short as 1000 kilometres, but often a jet stream extends to 10,000 kilometres. The width of a jet stream band can range from 400 to 1500 kilometres.

Jet stream wind speeds are usually stronger in winter than in summer because of the greater contrast between warm and cold air masses. The jet stream band will vary greatly in width. The wind speed peaks at the core of the jet and can reach speeds of over 400 kilometres per hour, but the maximums are generally in the 150 to 200 kilometre per hour range. The mean position of the jet stream also moves with the seasons. In winter, the strongest jet stream is usually positioned across southern Ontario. In summer, the strongest jet stream cores shift northward and may be located north of Ontario for short periods.

For most of us, the importance of the jet stream is how it relates to air travel. Aircraft flying in the same direction as the jet stream have their speed enhanced by the same amount as the wind at flight level. So an aircraft flying across the continent from west to east may have a flight time that is 1½ hours shorter than an equivalent flight from east to west. Airlines typically prefer to fly on routes that take the jet stream into account.

Winds

The downside of flying the jet stream is the turbulence that is often associated. Turbulent flow in the atmosphere occurs when winds change speed rapidly. Jet stream flow is not horizontally uniform, and the speed drops more quickly poleward of the jet maximum than equatorward. As well, the wind drops quite quickly into the stratosphere above the jet core. So flying the jet stream is a balancing act between travel time and turbulence, and commercial airline pilots are constantly monitoring the locations of jet streams and associated areas of clear air turbulence.

Nocturnal Jets

At night as the ground surface cools, so does the adjacent air. The air above the surface layer does not cool as quickly, so a warmer layer of air typically develops above the colder surface air. This situation is the reverse of the normal change of temperature with height, in which air cools with elevation. The reversed lapse rate, which displays warming with height, is known as an inversion. Nocturnal inversion conditions occur most nights. This thermal structure in the lower atmosphere often induces weakening of the wind adjacent to the ground surface, so the winds we experience at ground level are light overnight. At the top of the nocturnal inversion, the winds reach a maximum, which is known as a nocturnal jet. Maximum wind speeds can reach over 60 kilometres per hour. The depth of this nocturnal jet is only 100 metres or so, but because the nocturnal inversion is widespread, the nocturnal jet may exist as a "sheet" of strong winds several hundreds of kilometres wide and up to 1000 kilometres long. This jet breaks up in the morning as the nocturnal inversion

Winds from Around the World

- **Brickfielder:** a strong, dry, dust-laden summer wind from the desert in southern Australia.
- **Haboob:** a strong wind that occurs in the Middle East and along the southern edges of the Sahara in the Sudan; it is associated with extreme sandstorms and is most common in summer.
- **Levanter:** a strong easterly wind in the Mediterranean, especially the Strait of Gibraltar; it is associated with foggy, overcast or rainy weather, especially in winter.
- **Nor'easter:** a strong northeast wind that blows across the Atlantic provinces and the coast of New England.
- **Pampero:** a strong west or southwest wind in southern Argentina that brings with it cold polar air and is often associated with severe thunderstorms.
- **Santa Ana:** a strong, hot, dry wind that blows from the southern California desert through the Santa Ana Pass; most common in spring and autumn; can form at any time.
- **Simoom:** means "the poisoner"; a strong, dust-laden, cyclonic wind that blows across the Sahara Desert and the Arabian Peninsula, including Jordan and Syria; brings temperatures in excess of 54° C.
- **Sirrocco:** a hot wind in southern Spain that originates in the Sahara Desert.
- **Taku:** a powerful northeasterly wind that can occur in Alaska between October and April; wind speeds can reach hurricane levels.
- **Wreckhouse:** southeasterly gap winds funneled through the valleys of the Long Range Mountains in southwestern Newfoundland with downslope wind gusts to 200 km/h when they reach the flat plains flanking the mountains to the west. Most common during winter and spring with intense Atlantic storms.

Winds

dissipates with heating. On occasion, the jet becomes so intense overnight that the stability of the atmosphere breaks down. In this case, surface winds may become gusty under clear sky conditions, usually accompanied by a temperature increase as the warm air aloft mixes to the surface. Usually the gusty conditions are not prolonged, and the lower atmosphere repeats the inversion formation process. I have observed this phenomenon while working overnight as a shift forecaster—winds suddenly increased at a few locations for no apparent reason (i.e., no fronts or thunderstorms in the vicinity). It can be quite unsettling when the tranquillity of a clear, calm night is shattered.

Convection-Induced Winds

Cumulus clouds are formed when air near the ground is heated and begins to rise buoyantly in features known as thermals or cells. As the parcels rise, other air parcels move downward to replace the rising air. These parcels mix with the air at lower levels, resulting in a turbulent layer of air below the cloud base. An observer on the ground would feel the movement of this turbulent air as a gusty wind.

Smiths Falls downburst damage

> *Convection starts in the lowest layer of the atmosphere, where warm air rises. As it moves higher, the air expands and cools. Nearby air sinks back toward the ground to replace the rising air.*

As cumulus clouds develop vertically, precipitation may start to form within the cloud. The raindrops start to fall relative to the uplifting air, and as the drops grow, their downward velocity increases until it is faster than the upward moving air. This rain falls through the cloud, and when it exits the saturated base, evaporation starts. The evaporation process extracts heat from the surrounding air and causes cooling. The air within the rain shaft may, in fact, become colder than surrounding air, causing an accelerated downward push. The downward rushing rain mass also drags the air immediately adjacent to the droplets so that an extensive slug of cold, moist air accelerates toward the ground. Upon reaching the underlying land surface, this downdraft spreads out laterally, pushing dryer air ahead. At the ground surface, we observe this movement as a gust front. The gust front may be located well downwind of the precipitating shower cell, and it can appear in any direction from the cloud. When a gust front associated with a precipitating rain cloud passes, there is sometimes a sudden wind shift with no precipitation.

Winds

Thunderstorm Producing Plow Winds

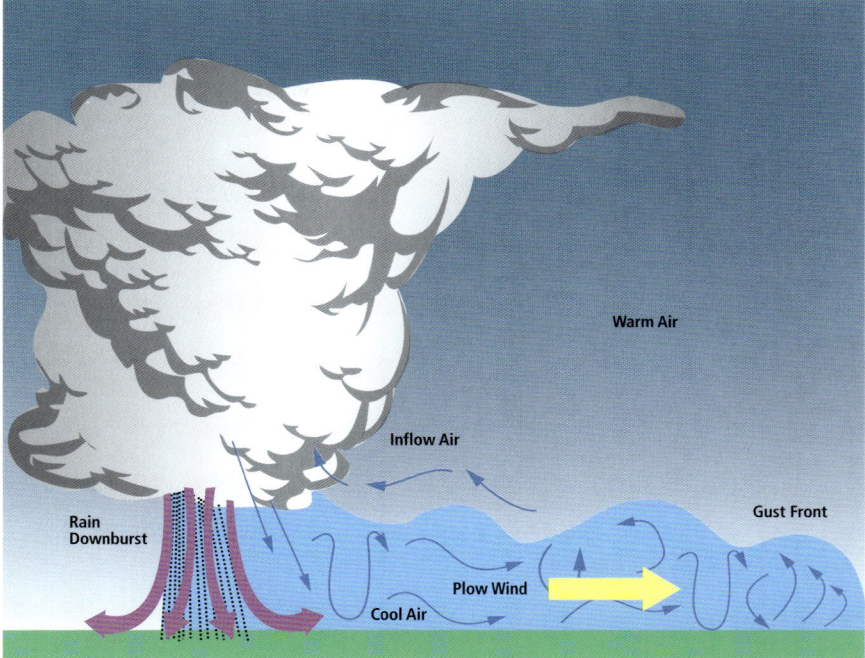

Fig. 5-6 Downward-rushing cold air moves rapidly along the ground, producing a plow wind close to the base of the storm and a gust front.

Most often, however, the wind shift is soon followed by rain.

This gust front has a number of impacts. The cold air can inhibit further convection near the raining parent cloud by choking off updrafts. The gust front can also undercut warm, moist air that is moving toward the parent cloud. Occasionally, the uplifted, moist air at the gust front starts a new line of convective showers. Intersecting gust fronts from thunderstorms with vigorous rain can result in the formation of a new rain cell, which may in turn grow into a mature thunderstorm.

The downburst from an actively precipitating thunderstorm occasionally results in strong, divergent winds that may damage trees or structures in their path. Wind speeds in severe downbursts can exceed speeds of 250 kilometres per hour, the same speeds that are observed in F3 tornadoes. In Canada, these destructive downbursts are often called plow winds because they cause damage that is generally in a straight line, generating a destruction path that is akin to the passage of a plow in soil. A downburst is classified as either a macroburst or a microburst. A microburst is less than four kilometres across and is typically more intense than the larger macroburst.

Winds

F2 tornado June 25, 2009. Note the circulating debris in the convergent tornadic winds. This shrapnel-like debris is lethal.

Tornadoes

The winds associated with tornadoes are the strongest in the atmosphere. Conventional wind measurement instrumentation cannot withstand the force of these winds, so the strength is usually an estimate based on the type of damage that occurs. The speed scale is named after Theodore Fujita, the research scientist who developed it in the 1960s. The so-called F-scale describes wind speeds in six ranges, from F0, with speeds up to 110 kilometres per hour, to F5 tornadoes, with wind speeds estimated over 420 kilometres per hour. In Ontario, few tornadoes have speeds above the F3 level—the Violet Hill tornado of April 20th, 1996, was estimated at F3—but the large tornadoes associated with the Barrie storm on May 31, 1985, reached the F4 level, with wind speeds of 330 to 410 kilometres per hour.

The tornado formation process is still not fully understood, but a credible theory is emerging based on observations of the tornado storm environment together with the understanding of thunderstorm dynamics. Most thunderstorms go through a normal lifecycle, from formation to decline, within about one hour. A few cells, known as supercells, last for several hours and can track several hundreds of kilometres. Some supercells develop strong mesocyclones, with which the stronger tornadoes are associated.

Supercells require a number of atmospheric conditions to come together in the correct sequence and at the correct time and location. Most important is the availability of a stream of warm, humid air in the lower atmosphere and dry, relatively cool air in the middle to upper atmosphere. This temperature condition gives rise to what is known as convective

Winds

instability. With a convectively unstable atmosphere, air parcels that rise from near ground level are warmer than their surrounding environment and continue to surge upward. For a supercell to form, strong solar heating is required, but a layer of warmer air above the surface that results in a thermal inversion should be present to suppress the convection until later in the day. Essentially, the inversion acts like a lid on a boiling pot of water—when the steam builds up, the lid blows off. Similarly, in the atmosphere, the inversion inhibits the early formation of cumulus clouds and small thunderstorms that would dissipate much of the energy.

Movement of colder air aloft over the warm, moist, lower-level air is favourable to the formation of the intense updrafts and downdrafts associated with thunderstorms. Such a condition is usually associated with an advancing cold front. Ahead of the cold front, an upper level jet stream adds an additional stimulus for thunderstorm development. Portions of the jet stream have areas of ascent that can help to further destabilize the upper atmosphere. As the thunderstorm cell moves forward, low-level winds feed moisture into the cell below its base. This moist air condenses, releases latent heat and creates air parcels that can accelerate upward at speeds of more than 30 metres per second in updrafts. Downdrafts composed of down-rushing, cold, precipitation-laden air can have speeds in excess of 10 metres per second. These

Fujita Scale Rating the Severity of Tornadoes

The Fujita scale is used to rate the severity of tornadoes as a measure of the damage they cause.

INTENSITY	ESTIMATED WIND SPEED	DAMAGE
F0	light winds of 60–110 km/h	some damage to chimneys, TV antennas, roof shingles, trees, signs and windows
F1	moderate winds of 120–170 km/h	cars overturned, carports destroyed and trees uprooted
F2	considerable winds of 180–240 km/h	sheds and outbuildings demolished, roofs blown off homes, and mobile homes overturned
F3	severe winds of 250–320 km/h	exterior walls and roofs blown off homes, metal buildings collapsed or severely damaged, and forests and farmland flattened
F4	devastating winds of 330–410 km/h	few walls, if any, left standing in well-built homes; large steel and concrete missiles thrown great distances
F5	incredible winds of 420–510 km/h	homes levelled or carried great distances, tremendous damage to large structures such as schools and motels, and exterior walls and roofs can be torn off

Fig. 5-7

Winds

updrafts and downdrafts are often adjacent to each other, and the zone between them is subjected to a high wind shear. This wind shear can impart a horizontal rolling action to the atmosphere just above the ground, near the cloud. With the assistance of the thunderstorm updraft, this rolling wind tube sometimes becomes vertically tilted, and if the rotation is cyclonic (counter-clockwise), it may create a cyclonic circulation called a low-level mesocyclone. Similarly, wind shear through the depth of the atmosphere can be ingested into the updraft, also causing it to rotate with time. This mid-level mesocyclone defines the supercell.

As the low-level mesocyclone continues its movement, ingesting warm, moist air, the rotation is stretched upward by the updraft, and links with the mid-level mesocyclone. Continued stretching of the rotating tube by the intense updraft further enhances its rotation speed, much like how figure skaters increase their rotation speed by pulling their arms closer to their body. The downdraft wraps around the updraft and helps to control the radius of contact of "the skates on the ice," further influencing the rotational speed of the low-level mesocyclone. If the downdraft is not too cold, yet still manages to pinch the "skater's" point of contact to a small radius, the spinning column can increase its rotation to reach damaging speeds, thus becoming a tornado. The most severe tornadoes often have even smaller spinning columns that move around the main column. These are known as suction vortices and frequently contain the strongest winds. There are a number of ways to form a tornado, and this is one that is often associated with supercell storms. But the science of meteorology continues to advance.

The lake breezes in Ontario not only focus the convection as previously described but also provide some of the wind shear that can be used to spin the updraft in a thunderstorm. These lake breezes, as well as geography, define the placement and orientation of Ontario's tornado alleys. It is no simple coincidence that the lightning hotspots of Ontario coincide exactly with the start of the province's tornado alleys.

Formation of a Tornado

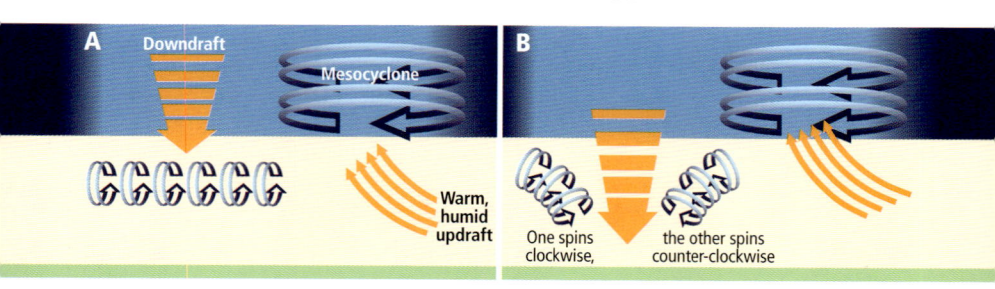

Fig. 5-8
A - low-level wind shear forms a spinning tube of air
B - the storm's downdraft pushes down on the spinning tube, tilting it into two columns

Winds

Surprisingly, it is not windy just a short distance from a tornado; the wind can even be calm with the supercell updraft overhead. Within the tornado itself, the violent winds turn any debris into deadly shrapnel. The damage path is a convergent trail of twisted debris, in contrast to the divergent pattern associated with a downburst. Typically, the debris falls from the northeast to northwest of the tornado. The sound of the tornado has been likened to the roar of a train or jet engine and results from the terrible damage being inflicted on the landscape. In contrast, the sound of severe hail is a softer and steady sound of the large stones hitting the ground or colliding in the air.

Waterspouts

Tornadoes associated with supercells that happen to cross a lake are sometimes referred to as waterspouts. Indeed, true waterspouts are rotating columns of air similar in appearance to that of the tornado. However, waterspouts can be created by intense instability that forms when cold air crosses warm water and the resulting instability lifts a vortex and stretches it in the vertical. Surface winds associated with these waterspouts have been observed to reach 165 kilometres per hour (F1). A waterspout can be associated even with a developing cumulus cloud. The maximum frequency for Ontario waterspouts is actually during cold outbreaks in autumn, when the lakes are still warm and the resulting low level instability is intense. Therefore, it is best to call a tornado crossing a lake a tornado instead of a waterspout. They may look similar, but a tornado is associated with stronger damaging winds and a supercell thunderstorm.

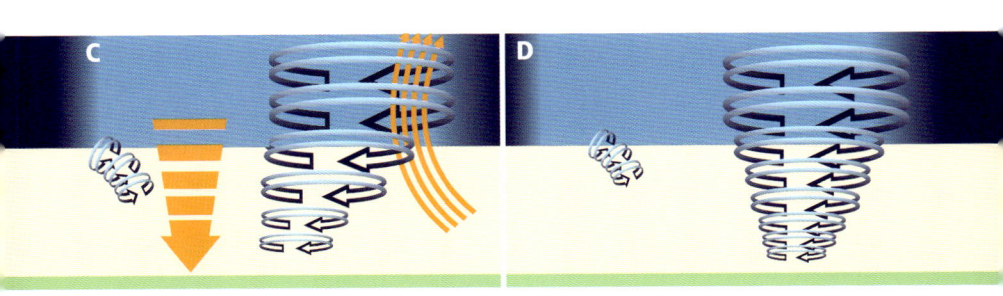

C - the counter-clockwise column is stretched into a tornado by the updraft
D - the column spinning clockwise, unstretched, spins slower. Sometimes it circles the tornado as a small funnel cloud.

Winds

Dust Devils

Dust devils are basically weak cousins of tornadoes. Unlike tornadoes they are not associated with any cloud. Dust devils are formed during late spring and summer on nearly calm, sunny days. Dark surfaces effectively absorb the sun's energy, and the surface in turn heats the adjacent air. The hot air rises rapidly in a column above the hottest area of the terrain, and hot air swirls in at the base of the column, so that the whole column takes on either a clockwise or counter-clockwise rotation. Dust devils are made visible by the dust, sand or debris that they pick up from the ground. The most vigorous dust devils can extend to heights approaching 1 kilometre, though most are only a few metres in depth. Surface wind strength can be up to 70 kilometres per hour. Dust devils are usually short lived and move slowly in the direction of the mean surface wind. The dark soils and flat terrain of the Holland Marsh in southern Ontario are especially conducive to spring dust devils.

Winds

Elevated Funnels (Cold Air Mass Funnels)

Elevated funnels typically associated with mid level, unstable conditions are common with cold low weather patterns. These elevated funnel clouds rarely reach the earth's surface. If wind damage does touch ground, they by definition become tornadoes but are short lived. Their strength is usually in the F0 category, and they cause little damage, if any. In Ontario, vigorous developing convective cells can create elevated funnels during any of the warm seasons, but the funnels are most common in spring and fall. The strongest portion of the updraft is aloft, and that is where the strongest tube stretching and visible funnel are observed. The circulation certainly extends to the ground but is typically far too weak to be noticed.

Winds

Windstorms and Tornadoes (from Environment Canada)

Tornado Watches

- Issued when conditions are favourable for the development of severe thunderstorms with the potential to produce tornadoes
- Lead time up to 2 hours, but usually much shorter
- There is time for preparation (secure loose items outdoors, notify others)
- Inform public to watch the sky for developments

Tornado Warnings

- Issued when a tornado is expected to develop soon or a tornado is nearby and will move into the area
- Lead time up to 2 hours, but usually much shorter
- Take cover

Windstorm and Tornado Safety

Outdoors

- Do not chase storms or tornadoes—they can change direction unexpectedly, and the shrapnel is lethal
- Get out of the storm path by moving at right angles to the direction of the storm's motion
- Get out of cars, mobile homes or campers and go to a sturdy, permanent building
- If driving, special care and attention is required
  - obey traffic laws
  - be wary of other drivers who may panic
  - do not seek shelter under a road overpass

Indoors

- Go to the lower floors, such as a basement under stairs
- Put as many walls as possible between you and outdoors
- Avoid windows, and never waste time closing windows
- Washrooms, hallways, stairwells and reinforced walls are best
- Avoid large open-span roofed areas such as gyms
- Hide under heavy desks, tables or workbenches with protective pillows or cushions

Fig. 5-9

Winds

"Windy Island" by Phil Chadwick

Onshore Winds and Shoreline Convergence

Strong winds are prevalent in many of the onshore areas of the Great Lakes. Winds increase to their full potential over the inland seas because of the lack of friction. In autumn and early winter, the instability created by the warm water after a summer of heating allows convective mixing of stronger winds aloft down to the surface. The winds become even stronger when they converge in the onshore areas and are forced to climb the terrain. The most damaging winds occur with and immediately after the passage of a cold front. These events are common each autumn on the eastern shores of the Great Lakes, most notably on the eastern shores of Georgian Bay and Lake Superior. Vegetation in these areas is flagged, reflecting the strength of the onshore winds. Strong winds are less likely in spring because the cold water has a stabilizing effect on the overlying air masses.

In Greek mythology, Zephyrus was the personification of the west wind, so a zephyr is a gentle breeze.

Chapter 6: Optics and Acoustics

The atmosphere consists of a mixture of gases, which have optical and acoustic properties and can produce illusions as we observe and listen. To understand the optical effects, it is first necessary to appreciate that the electromagnetic radiation emitted by the sun extends over a wide range or spectrum of wavelengths. The solar spectrum starts with the shortest and most energetic waves, known as gamma and x-rays, and then extends into the ultraviolet wavelengths. Visible light is in the middle of the spectrum, and infrared radiation, which we sense as heat, has the longest wavelengths.

Blue Sky

Why is the sky blue? This question was answered in the 19th century when scientists realized that gas molecules scatter electromagnetic radiation. As sunlight enters the top of the atmosphere, the gases immediately start scattering some waves in the light spectrum. The size of the gas molecules determines which wavelengths are scattered. The atmosphere consists of mostly oxygen and nitrogen, and the molecules of these gases are the right size to scatter blue light while leaving the longer wavelengths basically untouched. The blue

Optics and Acoustics

light rays bounce off other molecules as well. This scattering occurs in all directions—back toward the sun, at right angles and even at small angles to the solar rays—and gives the sky its apparent blue colour from the horizon to the sun. If one were up at the top of the atmosphere where there is little oxygen or nitrogen, the sky above would appear black—there is nothing to scatter light. As one progresses toward the earth's surface, the atmosphere takes on a dark blue colour as some blue light is scattered. Indeed, at the tops of mountains, the sky generally has a darker blue colour.

Water vapour molecules are large enough to scatter all the visible wavelengths, so the presence of water vapour results in a whitish haze when you look toward the horizon. Clouds consist of large water droplets that scatter all wavelengths, making the clouds appear white. Atmospheric pollutants consist of gases, each of which has its own characteristic molecular makeup and size. These gases have an ability to scatter back incident light of certain wavelengths. The pollutants that are responsible for what is known as photochemical smog scatter light back in the yellow part of the spectrum. That's

Why the Sky is Blue

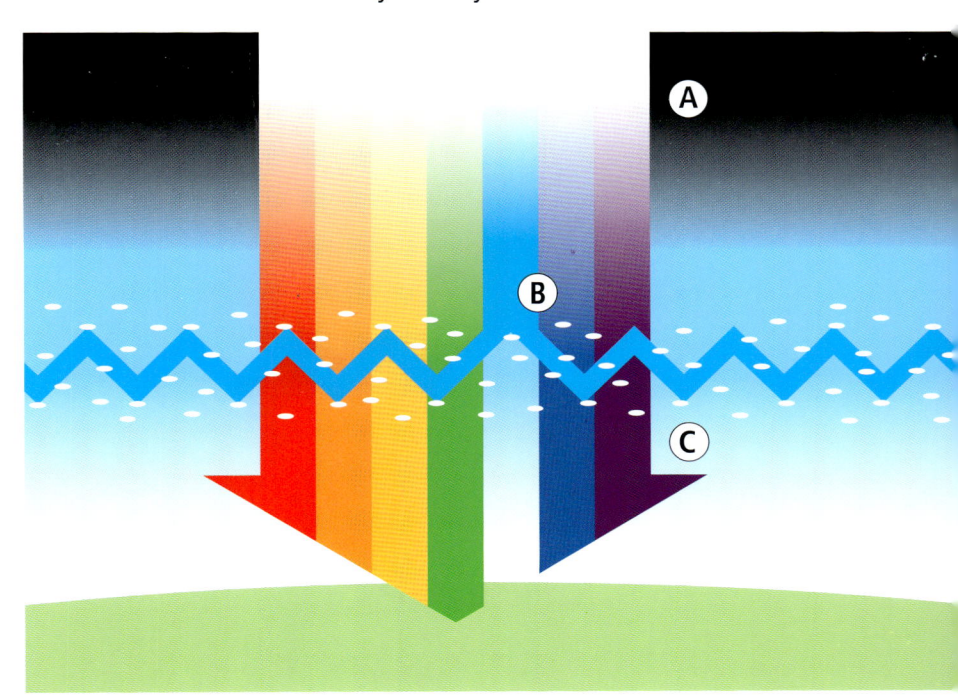

Fig. 6-1
A - white sunlight at the top of the atmosphere is composed of the full colour spectrum
B - some light in the blue colour spectrum is scattered in all directions by gas molecules in the atmosphere. As a result, we see blue sky in all directions.
C - the remaining light in other frequencies travels to the bottom of the atmosphere

Optics and Acoustics

why the elements of smog appear to have a yellowish tint. These human-made, or anthropogenic, pollutants are more evident around the major cities in Ontario. In summer and winter downwind of the larger cities, there is a brownish to orange bank of pollution that originates from the city's vehicles and industrial emissions.

Rainbows

Falling water droplets characteristically have a spherical shape. When a light ray passes through a water droplet, it reflects off the back of the inside surface and then passes back out the front. Each time the ray moves between the air and water surface, some refraction occurs and, like the action of a prism, the white light spreads into its colour spectrum. A rainbow is the observed action of millions of droplets acting together. The angles between the sun, the raindrops and the observer are all well defined by optical theory, with the result that the rainbow is essentially an arc that appears when the sun is at the observer's back, shining on a mass of raindrops. So, the rainbow appears in the opposite direction of the sun when the sun is relatively low in the sky and is illuminating a falling mass of raindrops. If we were to observe a bank of raindrops from an airplane or mountaintop, we would see a complete circle.

> **Old weather saying:**
> *Rainbow in the morning gives you fair warning.*

Formation of Primary and Secondary Rainbows

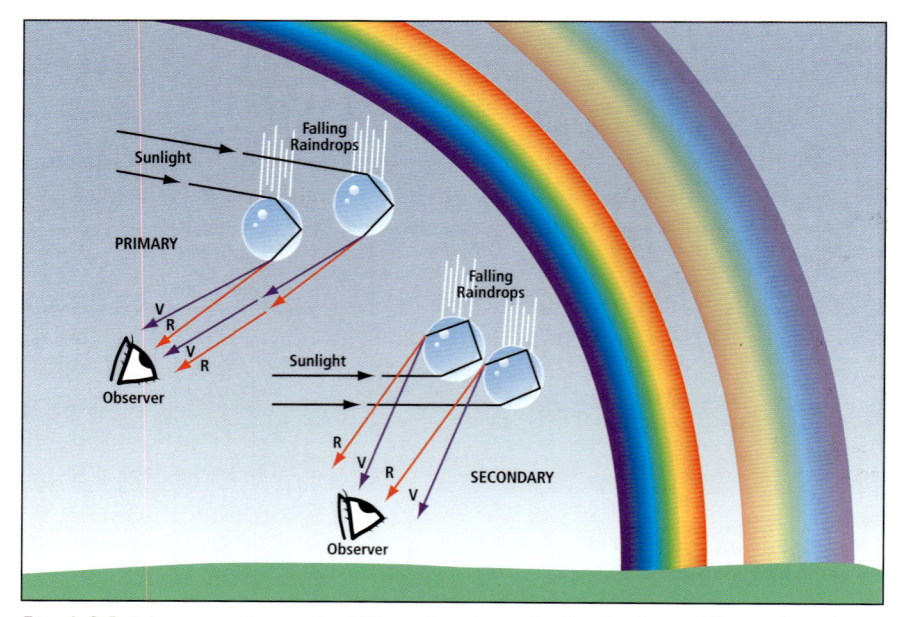

Fig. 6-2 Rainbows are the result of light refracting and reflecting from millions of raindrops.

Optics and Acoustics

Primary and secondary rainbows

What has been described above is known as the primary rainbow—the result of one reflection off the back of the rain droplet. Because the droplets are spherical, in reality multiple reflections occur inside each drop, giving rise to the appearance of a second or even third rainbow. With each reflection and refraction, there is some loss of brightness or intensity. As a result, the secondary bow is significantly less bright than the primary bow. This secondary bow is located outside the primary bow, and the colour sequence is reversed; instead of having red on the outside and blue on the inside, as do primary rainbows, the outside is blue and the inside is red.

In theory, multiple bows are possible, but because of increasing loss of intensity, they are not observed. Halley (of comet fame) actually calculated that the third rainbow arc, the tertiary bow, has an angular radius of about 320 degrees, and that it therefore should appear not in the part of the sky opposite to the sun, but as a circle around the sun itself. The observer would be standing in the rain, and the light from the sun would swamp the tertiary bow. This mathematical treatment of reflection and refraction is not conclusive in the real world, so readers are strongly encouraged to look for rainbows wherever they wish!

> *In the secondary rainbow, the colour scheme is reversed; instead of having red on the outside and blue on the inside, as do primary rainbows, the outside is blue and the inside is red.*

Optics and Acoustics

Sundog

Sundog Formation

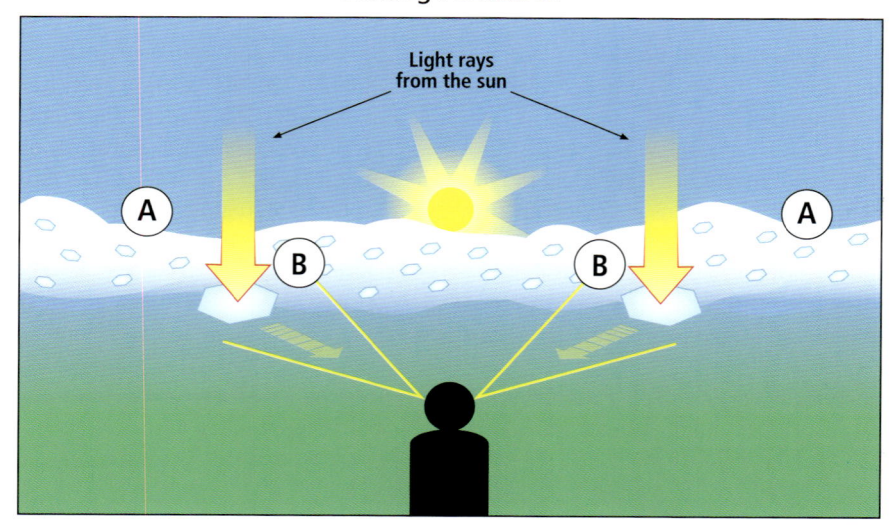

Fig. 6-3
A - cloud contains flat ice crystals
B - sun enters crystals and is bent at a 22 degree angle toward observer. Sundogs form on both sides of sun

Optics and Acoustics

Ice Crystal Phenomena

Several phenomena are a function of the unique characteristics of atmospheric ice crystals. More common in the cold season, these phenomena may be seen whenever ice is present in the clouds. The shape of the water molecule determines the shape of the resulting ice crystals. When formed in calm conditions, ice crystals are often six sided, or hexagonal. Sunlight can reflect off the ice crystal surfaces and can also pass through the crystal, which splits the white light in a prism-like fashion. This bending and splitting of light is known as refraction. Two commonly observed ice crystal phenomena result: sundogs and light pillars.

Sundogs (or, more correctly, parhelia) are observed when the sun passes through a layer of uniform ice crystals. Refraction of the sun's rays at the air-ice surface of each ice crystal produces the characteristic prismatic colour band. When the ice crystals lay uniformly in a horizontal plane, the refracted light is concentrated enough to form the sundogs, which can be seen at an angle of 22 degrees on either side of the sun. On cold winter days, a layer of ice crystals may form near ground level. These ice crystals are often called diamond dust because they reflect sunlight, giving the impression of glittering diamonds. On cold, calm winter mornings, the combination of diamond dust and ice crystals aloft gives the most spectacular displays of sundogs.

Light Pillar from Sunlight

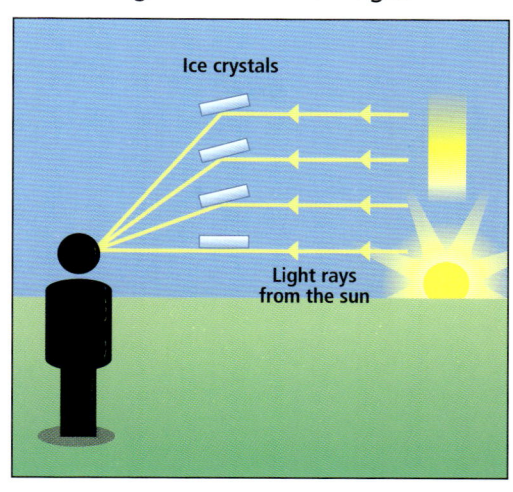

Fig. 6-4
Light reflected off underside of ice crystals results in a light pillar extending above the sun near the horizon

Light Pillar from Street Light

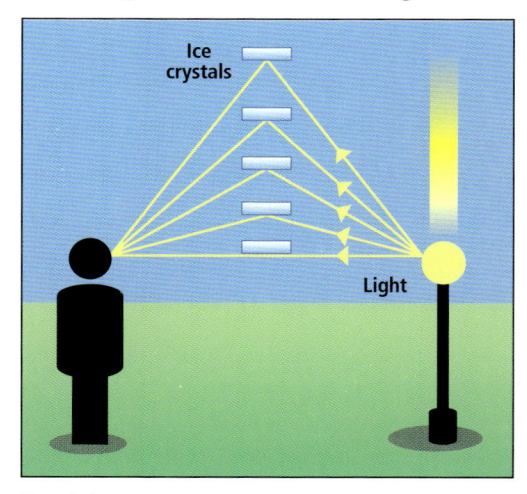

Fig. 6-5
Light reflected off ice crystals results in a light pillar extending above a street light

Optics and Acoustics

Light pillar extending above the sun

A **light pillar** is a reflection phenomenon that is formed in the presence of diamond dust (a ground-level cloud composed of ice crystals). The more or less uniformly oriented ice crystals act individually as tiny reflecting surfaces. As shown in Figures 6-4 and 6-5, the resulting effect is a shaft of light that seems to extend above and below the light source. Any relatively intense light source will produce light pillars. During cold, calm nights, streetlights often appear to have pillars extending vertically above them.

Optics and Acoustics

The Mirage Effect

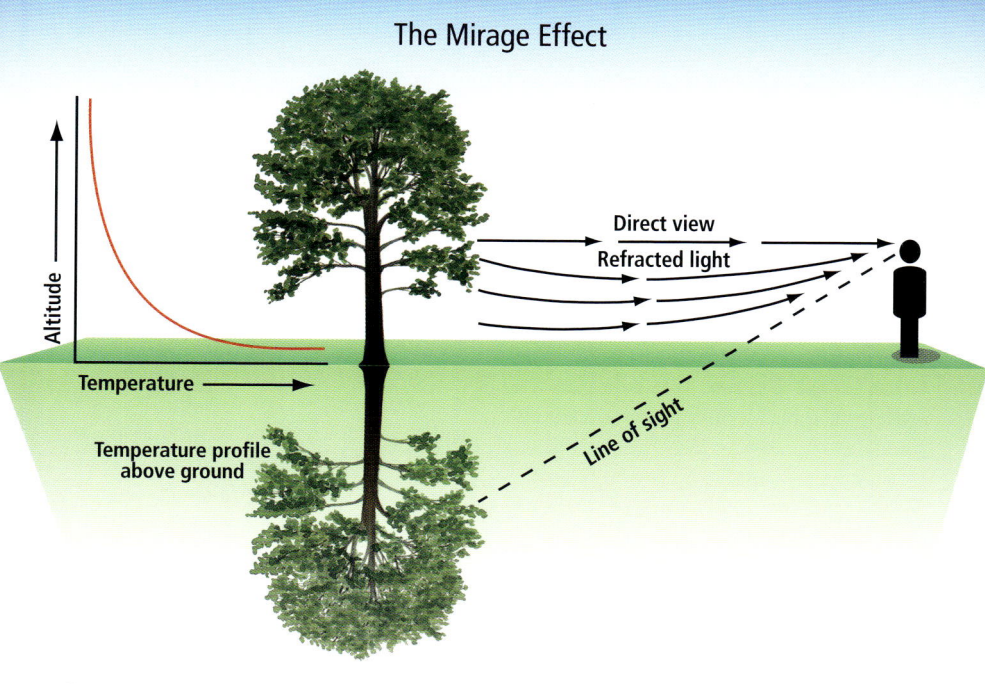

Fig. 6-6 Objects on the horizon may appear inverted.

Mirages

A mirage is a refraction phenomenon that occurs when light passes between layers of air with different densities. Light rays are bent away from the surface as they pass from colder air to intensely heated air close to the earth's surface. Such conditions happen when the sun heats sand or asphalt surfaces. Temperatures in the lowest metre or so above such surfaces can often vary up to 10° C, resulting in a significant air density difference between immediately adjacent layers of air. The light rays passing between the air layers bend, not in a pronounced way, but when we look toward the horizon, we often see an intense blue image, which is in reality the bending of light from above the horizon. Occasionally it is not the blue sky that we see but rather the image of an object on the horizon. In this case, the image is turned upside down and is usually distorted. Mirages are generally unstable because the extreme vertical temperature differences also cause the warmer, lighter air parcels at ground level to move upward and be replaced by denser, cooler air immediately above. The result is turbulence in the mirage layer, resulting in an unstable, shimmering effect.

Optics and Acoustics

Looming

This optical effect is similar to the mirage except that light rays are bent toward rather than away from the horizon. The condition happens when warm air is located above a colder layer near the earth's surface, which frequently occurs in the morning, when the air adjacent to the colder ground is overlain by warmer air aloft. This phenomenon is also common in spring over the cold Great Lakes. Because this event occurs most mornings, looming is a relatively frequent occurrence, usually manifested by an apparent tilting up of the horizon. As a result, at sunrise it can seem to observers as though they are in a broad, shallow bowl. This phenomenon also affects shorter wavelengths, which affects the behaviour of weather radar beams. Radars emit microwave radiation, which behaves much like light waves do. Weather radars emit their microwave beam at a slight angle above the horizon. However, during strong inversion conditions, the radar rays are bent toward the ground, even beyond the true horizon. The ground reflects these rays back along the same path through the atmosphere to the radar receiver. As a result, the radar display shows an image of ground clutter for some distance around the radar's location. In effect, the radar is "seeing" the bottom of the shallow bowl.

Crepuscular rays

Optics and Acoustics

Bright Water Horizons

The white line along the water horizon has long been painted by artists. When looking toward the sun across a water surface, both the surface area of the water that one is looking at and the angle of incidence to the water increase. The larger water surface area and increased reflection off the water combine to direct a large amount of light to the observer—the white line. This white line is intense and narrow for calm water and becomes increasingly broad and less intense with wave action or ripples that randomly tilt the planes of the reflecting "mirrors" of the water surface. The white line is also apparent with a rain shower approaching from the sunny side of the horizon. The rain dapples the water surface and results in countless tiny mirrors directing (or deflecting) light to your eye.

Crepuscular Rays

Crepuscular rays are also known as "Buddha's fingers," "sun drawing water" and the "ropes of Maui." They look like diverging beams of sunlight radiating outward from the sun. However, crepuscular rays are actually defined by the shadows cast by clouds, which offer sharp contrast with the sunlit columns in between. Crepuscular rays are not visible without clouds or some other obstacle present to cast atmospheric shadows. The light of the crepuscular ray is scattered to your eye by atmospheric molecules and dust. The divergent nature of these rays is simply a matter of perception in exactly the same way that train tracks appear to diverge toward you as you look along the tracks. Crepuscular rays are an everyday experience just waiting to be enjoyed.

Twinkling Stars

Stars typically twinkle during the approach of a warm front when strong winds in the upper atmosphere direct warm air poleward. The resultant turbulence in the warm air alters the path of the light from the stars, causing the stars to appear to flicker or even disappear from the perspective of the observer. The stars are very tiny sources of light, and altering the apparent position through atmospheric turbulence causes the star to jitter (or dance around) by distances greater than its small size. Planets are larger and do not twinkle. There is not enough atmospheric turbulence to significantly alter the apparent position of these larger celestial bodies. Stars do not twinkle in cold air masses because of the reduced amount of turbulence. As a result, if stars twinkle, you can expect that a warm front and inclement weather are on the way. If stars shine brightly without twinkling, look for cool, fair conditions.

Other Optical Effects

Because much of weather observation depends on watching the sky, people throughout the centuries realized that they could interpret what the weather would be like in the short range by observing sky and cloud conditions. Much of this observation became folklore, but some of it has an element of scientific rationale. An often-heard expression known to sailors is "red sky at night, sailor's delight; red sky at morning, sailor take warning." The explanation for this saying comes, in part, from the atmosphere's light-scattering capabilities, and from the direction of storm

Optics and Acoustics

motion at mid-latitudes. At sunrise and sunset, the sun's rays are travelling through a long atmospheric pathway, and the residual light is mostly red, giving cloud surfaces a reddish tint. At sunset, if there is no cloud in the western sky, clouds illuminated to the east look red. At mid-latitudes, storms move from west to east, and no cloud to the west indicates there are no approaching storms. So, "red sky at night [i.e., at sunset], sailor's delight" means no storms are on their way. However, if there are red clouds in the western sky at sunset, the clouds contain an abundance of moisture, and a wet storm is on the way. The saying could have a refinement that reads "red **clouds** at night, sailors take fright" (I made that up). If the sky is red in the morning, the rising sun is not obscured by cloud, and it may be illuminating the underside of a deck of cloud to the west. Again, because storms move from the west to east, it means potential storm clouds are approaching. Therefore, "red sky at morning [i.e., at sunrise], sailor take warning" means a storm is on its way.

People have also taken weather clues from nature, as can be seen in the expression "if bees stay at home, rain will soon come; if they fly away, fine will be the day." This expression can be explained by the fact that bees see in the ultraviolet spectrum and use polarized light, so they require sunny conditions to navigate to and from a food source. If it is cloudy, the light is no longer polarized and the bees stay close to home because they cannot navigate well.

Acoustical Effects

Like light waves, sound waves propagate through the atmosphere, and they have some similar behavioural properties that are notable under inversion conditions. The most frequently observed effect is that of reflection at discontinuities in the atmosphere. Sound waves are bent or refracted at relatively sharp discontinuities in the vertical thermal structure. At night, nocturnal temperature inversions form when air in contact with the earth's surface cools more quickly than the air aloft, resulting in the reverse of the normal situation in which temperature decreases slowly with height. In this inversion situation, there can be a sharp discontinuity in temperature. Sound waves propagating at an angle toward this discontinuity are refracted downward toward the earth's surface. This phenomenon is most readily apparent when the observer is relatively distant from a steady source of noise, such as a roadway with heavy traffic. At night, the noise from a distant highway, train or even general traffic from a city seems to be louder than it is during the day. Of course, other factors such as background noise levels and wind direction also come into play. The effect can be quite noticeable—for example, when one normally does not hear a freeway, but the sound becomes quite notable after sunset when the inversion becomes well established. A similar temperature inversion can be the result of a warm front.

Optics and Acoustics

Refraction of Sound Waves

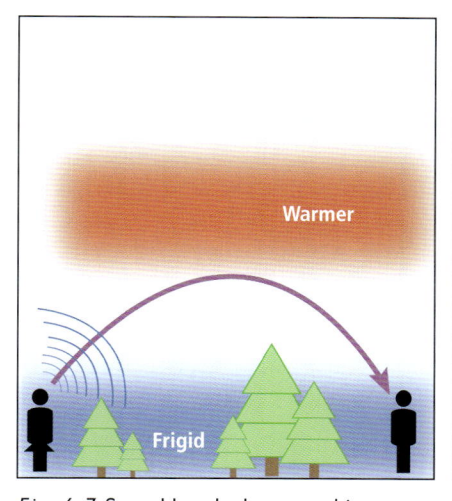

Fig. 6-7 Sound bends downward in a temperature inversion

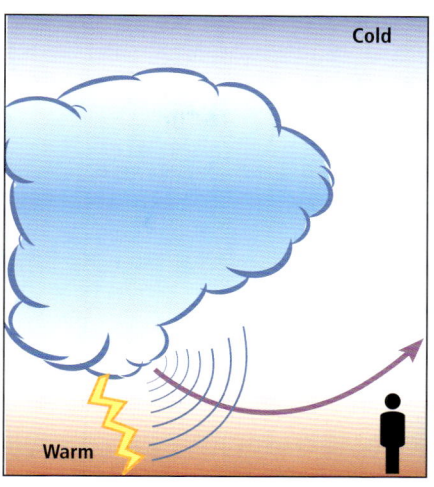

Sound bends upward if surface is warm

Sound waves can also be bent upward when temperatures decrease rapidly with height. For example, we often do not hear thunder from a distant lightning stroke because the sound wave is bent upward and away from us in an unstable atmosphere that cools rapidly with height. The sound from lightning strokes more than 20 kilometres away is almost never heard.

A simple acoustic effect relates to the rate of propagation of sound waves in air. Sound waves propagate at a speed of approximately 330 metres per second in the lower atmosphere. Light, on the other hand, is transmitted essentially instantaneously. We can use this difference to give a rough estimate of the distance to a lightning stroke by counting the number of seconds between seeing a lightning flash and hearing the crack of thunder. A pause of three seconds means the sound travelled about 1 kilometre, six seconds indicates 2 kilometres and so on. The timing of distant lightning strokes becomes more complicated because of reflection of the sound off atmospheric discontinuities, such as edges of clouds and shafts of falling rain.

In Canada's early days, 80 percent of the population was rural, and the people planned their activities by watching the sky and the weather. In the 2001 census, 80 percent of the population was urban.

Chapter 7: Ontario's Climate

Controls and Influences

The climate of a region is often described by statistics, such as averages, extremes and the most and least frequent occurrence of specific weather elements (precipitation, temperature, etc.). In reality, climatic averages are only lines on a graph. The day-to-day weather varies around these averages, from one weather condition to another. The climate defines what we can do comfortably, when we can do it, and what conditions may prevent certain activities. As the old saying goes, climate is what you may expect but weather is what you get.

Ontario's climate is diverse. Southern Ontario, generally considered to be the area south of Lake Nipissing, has a modified continental climate. Coniferous forests change to deciduous forests and farmland to the south. The rugged terrain of the Algonquin Highlands slopes southward to the Oak Ridges Moraine and the flatter land bordering the lower Great Lakes. The moderating effects of the Great Lakes make southern Ontario one of the mildest regions in Canada.

The northern portion of Ontario is in the Canadian Shield region and occupies 90 percent of the province. Most of the

Ontario's Climate

ONTARIO CLIMATE EXTREMES 1971–2000
(Adapted from the Meteorological Service of Canada data as posted on the Environment Canada website)

BARRIE
MAXIMUM (°C)	36.0 ON JULY 6, 1988
MINIMUM (°C)	-35.0 ON JANUARY 4, 1981
DAILY RAINFALL (MM)	96.0 ON JUNE 2, 1995
DAILY SNOWFALL (CM)	65.0 ON JANUARY 9, 1978
SNOW DEPTH (CM)	59.0 ON FEBRUARY 9, 2001

KENORA
MAXIMUM (°C)	35.8 ON JULY 14, 1983
MINIMUM (°C)	-43.9 ON JANUARY 20, 1943
DAILY RAINFALL (MM)	153.5 ON JULY 27, 1993
DAILY SNOWFALL (CM)	36.3 ON APRIL 10, 1957
SNOW DEPTH (CM)	145.0 ON MARCH 5, 1966

KINGSTON
MAXIMUM (°C)	34.3 ON JULY 15, 1983
MINIMUM (°C)	-34.5 ON JANUARY 4, 1981
DAILY RAINFALL (MM)	128.8 ON SEPTEMBER 14, 1979
DAILY SNOWFALL (CM)	46.2 ON DECEMBER 16, 1974
SNOW DEPTH (CM)	64.0 ON DECEMBER 22, 1977

LONDON
MAXIMUM (°C)	38.2 ON JUNE 25, 1988
MINIMUM (°C)	-31.7 ON JANUARY 24, 1970
DAILY RAINFALL (MM)	89.1 ON SEPTEMBER 29, 1986
DAILY SNOWFALL (CM)	57.0 ON DECEMBER 7, 1977
SNOW DEPTH (CM)	70.0 ON DECEMBER 10, 1977

NIAGARA FALLS
MAXIMUM (°C)	38.3 ON AUGUST 26, 1948
MINIMUM (°C)	-25.0 ON FEBRUARY 15, 1943
DAILY RAINFALL (MM)	95.3 ON AUGUST 24, 1954
DAILY SNOWFALL (CM)	45.7 ON FEBRUARY 19, 1940
SNOW DEPTH (CM)	47.0 ON MARCH 3, 1984

OTTAWA
MAXIMUM (°C)	37.8 ON JULY 4, 1913
MINIMUM (°C)	-38.9 ON DECEMBER 29, 1933
DAILY RAINFALL (MM)	93.2 ON SEPTEMBER 9, 1942
DAILY SNOWFALL (CM)	55.9 ON JANUARY 29, 1894
SNOW DEPTH (CM)	97.0 ON FEBRUARY 24, 1971

OWEN SOUND
MAXIMUM (°C)	35.0 ON JUNE 10, 2000
MINIMUM (°C)	-34.0 ON FEBRUARY 18, 1979
DAILY RAINFALL (MM)	75.7 ON AUGUST 19, 1968
DAILY SNOWFALL (CM)	62.0 ON DECEMBER 10, 1995
SNOW DEPTH (CM)	88.0 ON DECEMBER 12, 1995

SAULT STE. MARIE
MAXIMUM (°C)	36.8 ON JULY 7, 1988
MINIMUM (°C)	-38.9 ON JANUARY 23, 1948
DAILY RAINFALL (MM)	116.6 ON MAY 31, 1970
DAILY SNOWFALL (CM)	61.0 ON FEBRUARY 10, 1947
SNOW DEPTH (CM)	140.0 ON DECEMBER 12, 1995

SUDBURY
MAXIMUM (°C)	38.3 ON JULY 31, 1975
MINIMUM (°C)	-39.3 ON JANUARY 10, 1982
DAILY RAINFALL (MM)	112.0 ON SEPTEMBER 3, 1970
DAILY SNOWFALL (CM)	38.8 ON MARCH 10, 1992
SNOW DEPTH (CM)	145.0 ON MARCH 16, 1959

THUNDER BAY
MAXIMUM (°C)	40.3 ON AUGUST 7, 1983
MINIMUM (°C)	-41.1 ON JANUARY 30, 1951
DAILY RAINFALL (MM)	131.2 ON SEPTEMBER 8, 1977
DAILY SNOWFALL (CM)	51.6 ON JANUARY 20, 1956
SNOW DEPTH (CM)	179.0 ON JANUARY 22, 1956

TORONTO
MAXIMUM (°C)	40.6 ON JULY 8, 1936
MINIMUM (°C)	-32.8 ON JANUARY 10, 1859
DAILY RAINFALL (MM)	98.6 ON JULY 27, 1897
DAILY SNOWFALL (CM)	48.3 ON DECEMBER 11, 1944
SNOW DEPTH (CM)	65.0 ON JANUARY 15, 1999

WINDSOR
MAXIMUM (°C)	40.2 ON JUNE 25, 1988
MINIMUM (°C)	-29.1 ON JANUARY 19, 1994
DAILY RAINFALL (MM)	94.6 ON APRIL 20, 2000
DAILY SNOWFALL (CM)	36.8 ON FEBRUARY 25, 1965
SNOW DEPTH (CM)	42.0 ON FEBRUARY 9, 1982

Fig. 7-1

Ontario's Climate

land is forest, mixed with areas of swamp, muskeg and small lakes. The portions northwest of Lake Superior have a true continental climate like that of Manitoba, with large annual temperature swings and more precipitation in the summer months as a result of convection. The areas around Hudson Bay, James Bay and Lake Superior are moderated by the heat capacity of the surrounding waters and have more of a modified continental climate.

The seasons are distinctly different in Ontario, and most people have their favourite. Summers tend to be warm with periods of hot, humid, hazy weather. Precipitation comes haphazardly in the form of showers and thunderstorms. Many people prefer autumn for its warm, sunny days, lack of biting insects and forests brightened by the colourful deciduous trees. Winter means snowsqualls off the Great Lakes, temperatures bouncing above and below the freezing mark for the south, and cold and probably windy weather for the north. The snow and ice makes this a favourite time for many—and again there are no biting insects. Spring tends to be a cool, damp, cloudy season, but it doesn't last long because warm air frequently surges northward from the United States.

Air Masses

Ontario is subjected to an assortment of air masses, as illustrated in Figure 7-2. These air masses represent the characteristics of the regions in which they were formed. Ontario is mainly influenced by either arctic air masses or maritime tropical air masses. The tropical air masses that originate from the equatorial Atlantic or the Gulf of Mexico tend to be hot and humid—loaded with energy for storms. The arctic air masses are cooler. In winter, these arctic air masses originate from frozen tundra, so they are dry as well as extremely cold and are thus called continental arctic air masses. Sometimes in winter, arctic air passes over open oceans, such as the Gulf of Alaska, and is modified to become a maritime arctic air mass, which occasionally finds its way to Ontario. In summer, the Arctic is a large moisture source comprised of open lakes and marshes, so all summer arctic air masses are indeed "maritime." These maritime arctic air masses typically dominate northern Ontario but will sometimes push southward with the steering flow. The continental arctic air masses are only important during winter, but they, too, will move southward, bringing dry, cold conditions when the upper atmospheric winds are northerly. The maritime polar air mass and winter maritime arctic air mass must scale the Rocky Mountains on their occasional visits to Ontario. The Rockies wring out much of the Pacific moisture because the air must climb over the high terrain before pushing eastward toward Ontario as a much drier air mass.

Simply put, Ontario is a confrontation zone for air masses. The main summer air mass battles over the province are between the maritime arctic air arriving from the north and the maritime tropical air pushing in from the south. In winter, continental arctic air can battle it out with either maritime tropical or maritime polar air. The battles are usually very stormy whenever tropical air is involved.

Ontario's Climate

Winter Air Masses and Circulation

Continental Arctic
- very cold, -25 to -50° C
- dry, very stable
- pronounced temperature inversion

Maritime Arctic
- very unstable
- clouds, frequent showers or flurries
- visibility good except in showers

Maritime Polar
- milder and more stable than arctic air

Pacific Maritime Tropical
- light winds, cooler than Atlantic air
- comes to North America from west or northwest
- stable in lower 1000 m (marine stratum)

Atlantic Maritime Tropical
- comes to North America from south or southeast
- warm and humid

SST - Sea surface temperature

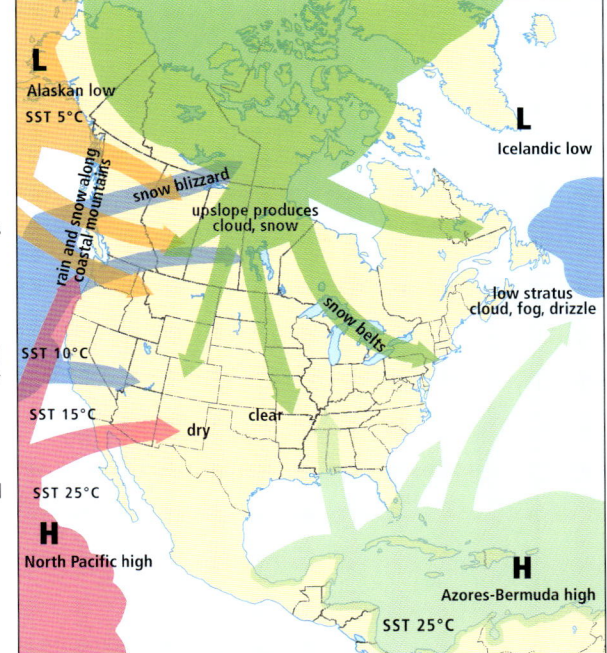

Map labels: Alaskan low, SST 5°C, L, Icelandic low, rain and snow along coastal mountains, snow blizzard, upslope produces cloud, snow, low stratus cloud, fog, drizzle, snow belts, SST 10°C, SST 15°C, dry, clear, SST 25°C, H, North Pacific high, H, Azores-Bermuda high, SST 25°C

Summer Air Masses and Circulation

Continental Tropical
- hot, dry, unstable

Maritime Arctic
- continental air modified by open seas, lakes and swamps

Maritime Polar
- warmer and more stable than maritime arctic air

Pacific Maritime Tropical
- high pressure blocks moist air

Atlantic Maritime Tropical
- oppressively hot and humid
- unstable, frequent thunderstorms

SST - Sea surface temperature

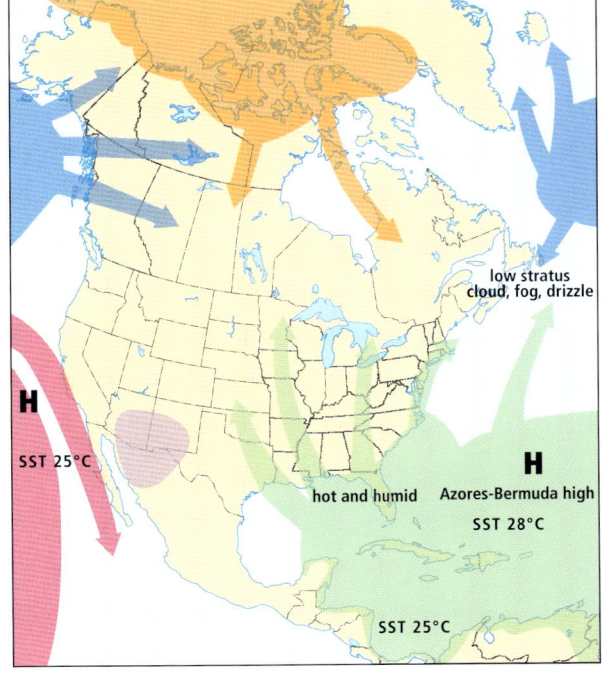

Map labels: low stratus cloud, fog, drizzle, H, SST 25°C, hot and humid, Azores-Bermuda high, SST 28°C, SST 25°C

Fig. 7-2

Ontario's Climate

Global Circulation Pattern

The global circulation pattern is the other major influence on Ontario's climate. Westerly winds in the upper atmosphere generally guide the motion of weather systems over the Ontario landscape. These weather systems provide much of the day-to-day variation in weather conditions. Frontal zones, which are the boundaries between the air masses, are part of these systems. Light rainfalls or showery conditions in the warm season are usually associated with frontal weather systems. Storms associated with fronts are stronger in winter than in summer and can bring about severe weather events.

El Niño / La Niña

The global atmospheric circulation pattern undergoes periodic changes, and El Niño of the tropical Pacific Ocean has a significant influence on the climate in Ontario as well as in other parts of the world. This phenomenon demonstrates the link between circulation patterns in the atmosphere and those in the oceans. The main currents of the Pacific basin are shown in Figure 7-3. The trade winds that blow from east to west along the equatorial Pacific drag surface ocean waters toward the west. Accompanying the westward drift of warm water is an upwelling of cool, deep, nutrient-rich water along the South American coast, which gives rise to a productive fishing industry off Peru and Ecuador. The trade

Pacific Ocean Currents

→ warm water current
→ cold water current

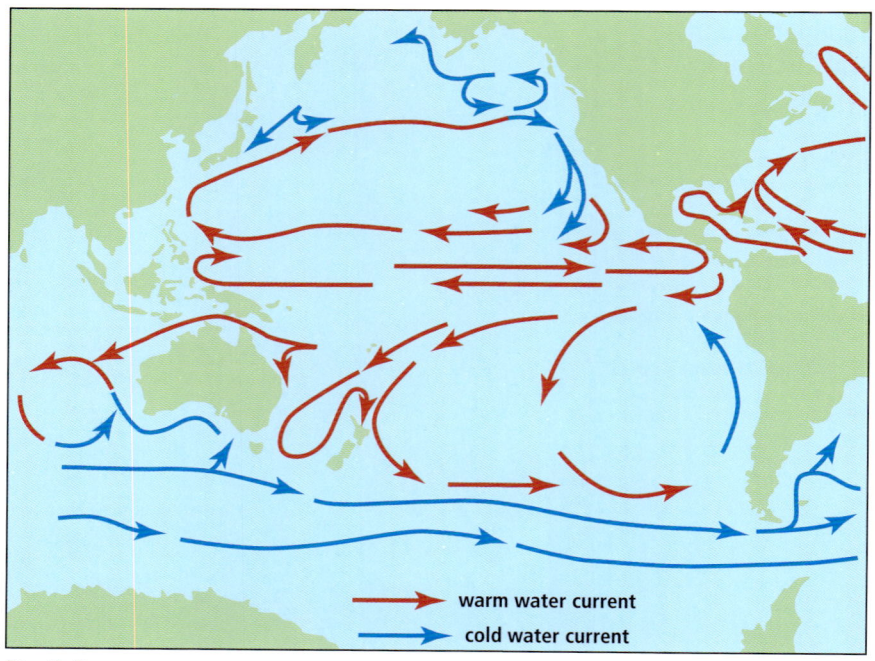

Fig. 7-3

Ontario's Climate

winds occasionally slacken or even reverse, and the accumulated warm water off the Indonesian coast begins to shift back across the Pacific, taking about two months to reach the coast of South America. When the bulge of warm water reaches the coast of South America, it cuts off the upwelling of the nutrient-rich waters. The warming ocean event usually starts in December, and the local fishermen named it El Niño, Spanish for "the little boy" and a reference to the Christ child. When the waters in the eastern Pacific are colder than normal, what is known as La Niña (or "the little girl") occurs. La Niña is the antithesis of El Niño.

El Niño warming events occur on an irregular basis at intervals of two to seven years. Accompanying this oceanic cycle is an oscillation in atmospheric pressure on either side of the Pacific Ocean. At the onset of El Niño, average pressures over the western Pacific are higher than normal

Typical January–March Weather Anomalies and Atmospheric Circulation During Moderate to Strong El Niño & La Niña

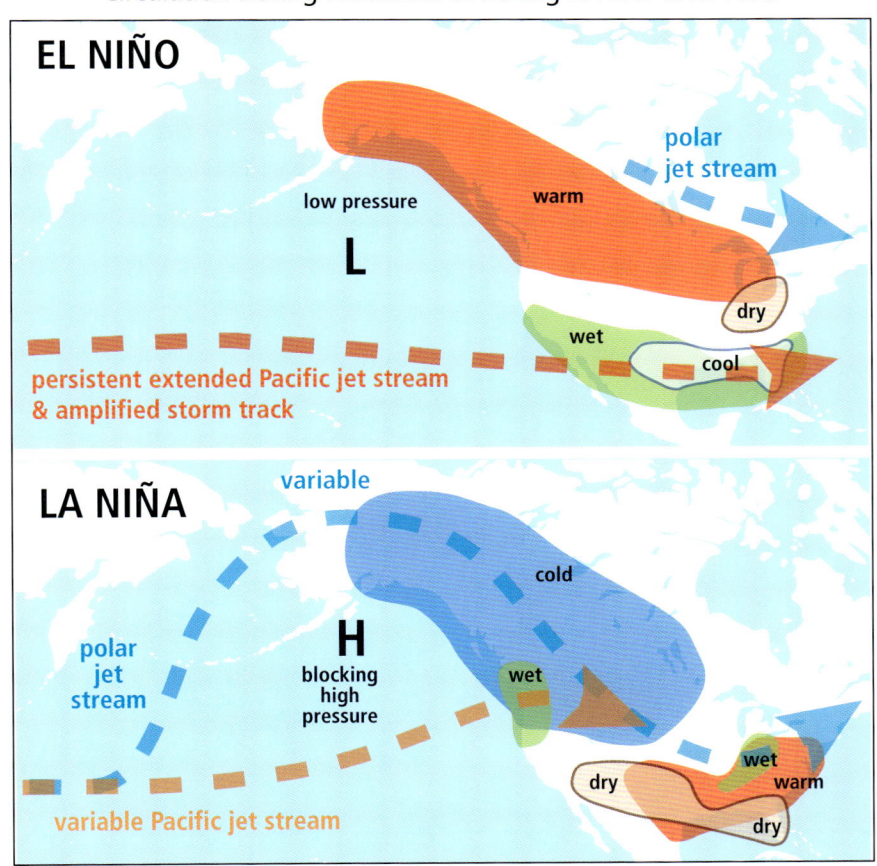

Fig. 7-4

117

Ontario's Climate

and are accompanied by lower than normal pressures in the east. This atmospheric seesaw cycle is known as the Southern Oscillation. The ocean and atmospheric cycles are linked and together are known as the ENSO or El Niño-Southern Oscillation.

Strong El Niño and La Niña episodes have a noticeable effect on the weather conditions in western North America. During an El Niño event, a stronger area of low pressure prevails over the west coast of North America, and the jet stream across the northern Pacific takes a southerly track, which directs warmer than normal air into western Canada. This creates an upper ridge in the atmospheric long wave pattern. The wave nature of the atmosphere then results in an upper trough downstream, over eastern Canada. The jet stream follows this pattern and takes a southeasterly track across the northern Great Lakes or even farther to the northeast. Therefore, during a strong El Niño event, winters in Ontario (especially western Ontario) are milder than average, with anomalies up to 3° C. There tends to be less precipitation, and snowsquall activity is weakened. This is not a good pattern for skiers. Southern Ontario is unlikely to have a white Christmas with El Niño. An El Niño event permits fairly accurate long-range winter forecasts. An El Niño summer tends to be hotter than normal with a higher probability of pulse type thunderstorms rather than supercell convection.

During a strong La Niña, the normal area of low pressure off the BC coast is replaced by higher pressure. A split in the jet stream pattern accompanies the

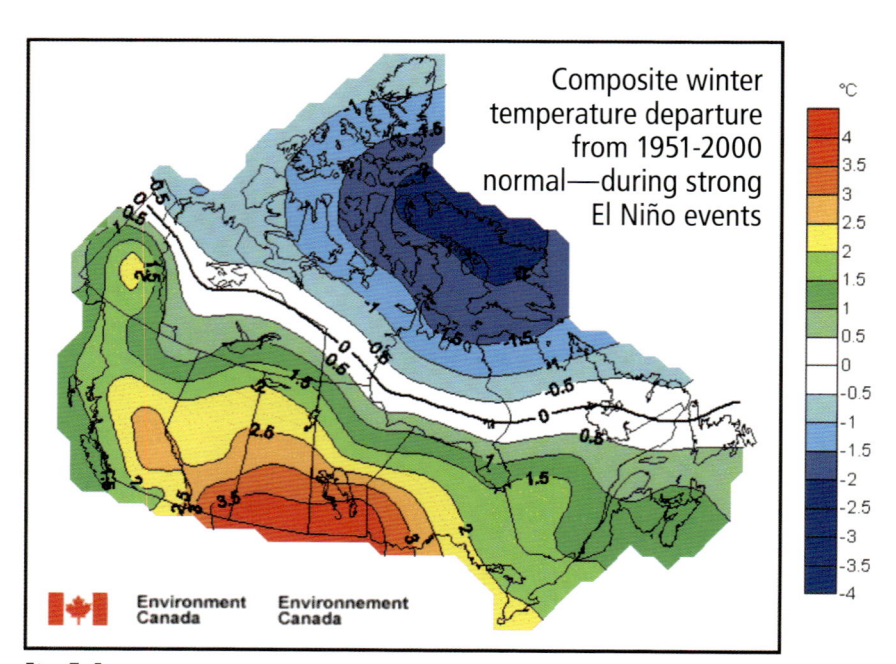

Fig. 7-5

Ontario's Climate

pressure change, and the northern portion of the jet stream moves over Alaska before heading southeastward along the Continental Divide, creating an upper trough in the atmospheric long wave pattern and an upper ridge downstream, over eastern Canada. The jet stream follows this pattern and takes a generally southwesterly path across southern Ontario. As a result, in a La Niña winter, Ontario temperatures tend to be below normal (especially over northern regions), and precipitation in southern parts of the province tends to be above normal. A La Niña summer tends to be cooler than normal with a higher probability of supercell convection because of the close proximity of the jet stream. La Niña is not as reliable a weather predictor as El Niño. The locations of the jet streams during these events are shown in Figure 7-4.

One of the strongest El Niño events occurred in the winter of 1982–83, when there were widespread impacts across the globe. It was dubbed the El Niño of the century. The event was very important for Alberta but less so for Ontario. The precipitation deficit for Alberta was 60 to 70 percent. For northwestern and southern Ontario, the precipitation deficit was only 20 to 30 percent. At the same time, northern Ontario received more than 50 percent more precipitation than normal. The storm track had simply shifted to the north. Seasonal maximum ice cover in the winter of 1982–83 for the combined area of the Great Lakes was the lowest (25 percent) since 1963.

The winter of 1997–98 was affected by another record-breaking El Niño event. The area around the Great Lakes experienced the warmest winter in 66 years

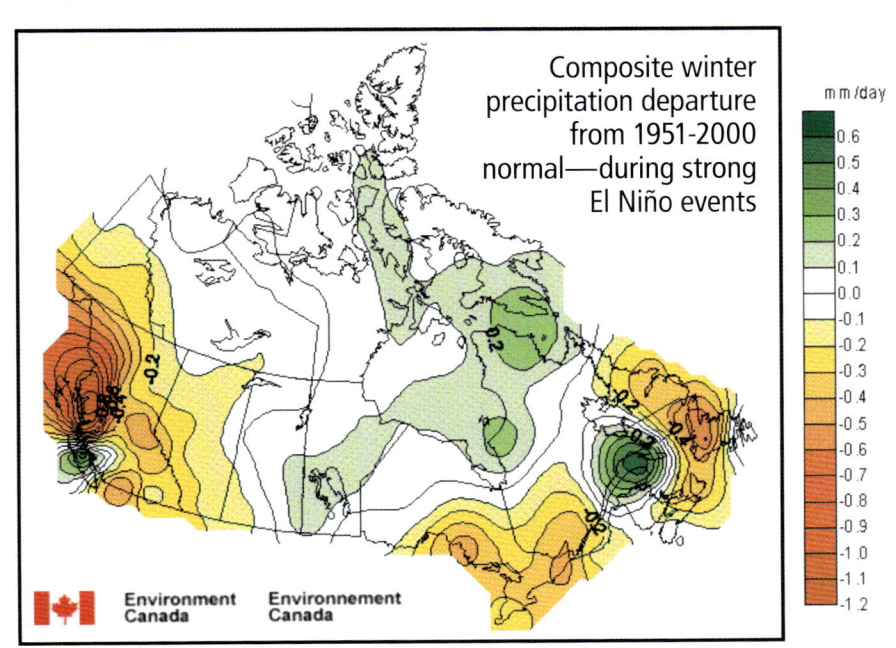

Fig. 7-6

Ontario's Climate

with temperatures 6° C above normal, and the seasonal maximum ice cover for the area set a new record low (15 percent), beating the record set by the 1982–83 El Niño. Toronto had the warmest February since records began in 1840. No snow fell east of the Ottawa valley. The ice wine industry of the Niagara area lost more than $10 million because temperatures did not reach the critical harvesting conditions of an extended period of -8° C, and non-migrating birds ate the crop. During the summer of 1998, precipitation was up to 40 percent below normal from the West Coast, through the Prairies and into southern Ontario. Both the Prairies and the farming communities in Ontario and Québec were hit by dry conditions caused by a shortfall of about 20 percent in the amount of precipitation.

An El Niño of moderate strength occurred in 2002–03. December 2002 was largely snow-free across western Canada, and central and southern Ontario received only 40 to 50 percent of their average precipitation, while areas north of Ontario received 80 percent or more additional precipitation.

The 1988–89 La Niña delivered frigid air to Ontario in February. The lower Great Lakes recorded their coldest February since 1980. Sudbury and Earlton were the coldest since 1979, and Kenora experienced the coldest February since 1967.

The 1995–96 La Niña brought heavy snow onshore off Lake Superior and Georgian Bay. These areas received 250 to 350 centimetres of snow through the active snowsquall period of November and December. Snowfalls across most of Ontario and northern Québec in January 1996 were up to double the average January amount.

A moderate La Niña in the winter of 2007–08 brought eastern Canada a winter of abnormally high snowfall. The snow arrived and never left, with new snowfalls every few days. Northern Ontario had temperatures averaging 4° C below normal. The south had average temperatures but way more snow than normal. Ottawa snowfall was 432.7 centimetres—just 12 centimetres short of the snow record of 444.6 centimetres recorded in 1970–71. These snowfall amounts are supposed to occur only once every 1000 years. Other major cities also came close to breaking all-time snow records—Toronto missed by just 13 centimetres, with 194 centimetres observed. The 70-year-old record still stands.

Terrain

The climate across the province is also determined by what is known as the physiography of the landscape. Ontario has four major regions: the Canadian Shield, the Hudson Bay Lowland, the Great Lakes Lowland and the St. Lawrence Lowland. The Canadian Shield covers two-thirds of Ontario and is made of igneous and metamorphic rocks that form flat plateaus and low hills. The Hudson Bay Lowland has flat sedimentary rocks. The Oak Ridges Moraine and the Niagara Escarpment are both part of the Great Lakes Lowland. They are subtle geographic features that have important influences on the local climate. The Niagara Escarpment is only a 335-metre-high ridge, and the Oak Ridges Moraine isn't even that high. The 160-kilometre-long Oak Ridges Moraine parallels the north

Ontario's Climate

shore of Lake Ontario and was formed 12,000 years ago by advancing and retreating glaciers. The low range of hills can dam cool air to the north in winter, favouring freezing or frozen winter precipitation. Summer lake breezes seldom make it north of the moraine. The Niagara Escarpment is a biosphere reserve stretching 725 kilometres from Lake Ontario (near Niagara Falls) to the tip of the Bruce Peninsula (between Georgian Bay and Lake Huron). Much of the Escarpment corridor is forested and crosses two major forest types: boreal needle leaf forests in the north and temperate broadleaf forests in the south.

High terrain modifies temperatures, especially at night. Ranges of hills can also affect the climatic character. The highlands west of the Niagara Escarpment are typically cooler at night than the lower terrain to the east. The base of the Niagara Escarpment near Lake Ontario creates an ideal milder microclimate for wine grape vineyards. The east-facing slopes of the escarpment are also the first to sprout cumulus clouds on a summer day. At night, colder air may pool along river valleys, resulting in earlier frosts than at nearby locations outside of the valleys. Peterborough, in the valley of the Otonabee River, is a well-known frost and fog hollow. South-facing slopes along the Oak Ridges Moraine are popular with farmers because the slopes get more direct sun and are therefore warmer and drier, especially early in the growing season.

The northern half of the province is dominated by the boreal plain and forest, which consists of a mixture of forested areas dotted with lakes and a few relatively significant areas of glacial remains.

The far north above the tree line is sub-arctic on the shores of James Bay and Hudson Bay, which is connected to the Arctic Ocean. The climate is harsh.

Oak Ridges Moraine and Niagara Escarpment

Fig. 7-7

Ontario's Climate

The Great Lakes

The Great Lakes have a huge impact on the climate of Ontario and on southwestern Ontario in particular. These impacts include the following:

- moderation of temperature (warmer autumns; cooler springs)
- intensification of storms, especially in autumn when the lakes are at their warmest
- increasing precipitation amounts and cloud cover with cold air masses mainly in autumn and winter; snowsqualls deliver 250 to 400 centimetres of snow to the onshore areas each winter.
- decreasing precipitation amounts and cloud cover with cool air masses mainly in spring and summer
- creation of fog with warm air masses in spring
- focussing of summer thunderstorms along lake breeze fronts (convergence lines)
- increasing wind speeds along the shoreline.

The convergence of the lake breeze fronts between Windsor and Sarnia, east of Lake St. Clair, is marked by an Ontario lightning hotspot. A second hotspot is located along a line from the southern tip of Georgian Bay to southeast of Barrie. This second lightning hotspot is the convergence zone of the lake breezes from the southern shore of Georgian Bay and off the north shore of Lake Ontario in an air mass with prevailing southwesterly breezes. The southwesterly downsloping off the Niagara Escarpment also gives this lightning hotspot a boost. Two highland areas in southern Ontario—Algonquin Park (northwest of Ottawa) and the Dundalk Highlands (southwest of Barrie)—experience lightning much less frequently than the lowland areas surrounding them.

> **The Great Lakes sailor is wild-ocean nurtured; as much of an audacious mariner as any.**
> —Herman Melville, *Moby Dick*

Lightning Hotspots

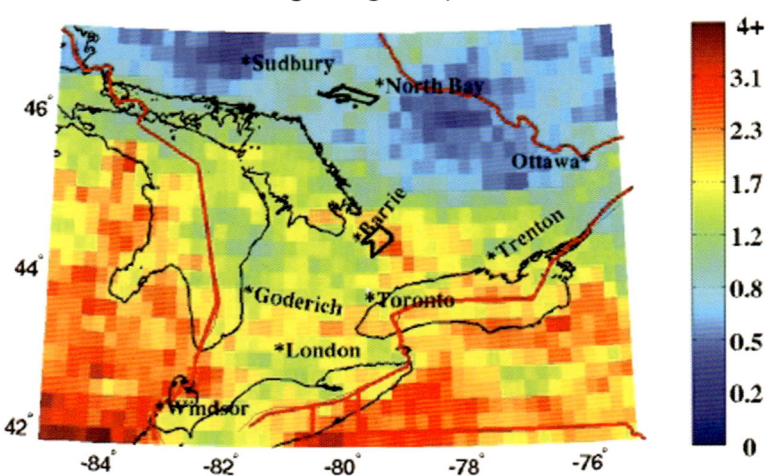

Fig. 7-8

Ontario's Climate

Temperature

Temperature is the weather element that has the greatest influence on living entities. Temperature variations largely control the viability of plants, and temperature means and extremes define human comfort levels. In Ontario, the temperature pattern across the province is an indication of the variation of the sun's angle from the south to the north. The Great Lakes only occupy the southern half of the province, and that is where the moderating effects are most noticed.

Annual mean temperatures are warmest in the south and drop off at a fairly uniform rate north of the Great Lakes toward Hudson Bay.

The lowest Ontario temperature on record is -58.3° C, measured at Iroquois Falls on January 23, 1935.

Apparent Temperature

Example: If temperature is 35° C and relative humidity is 50%, the apparent temperature is on the edge of very hot range.

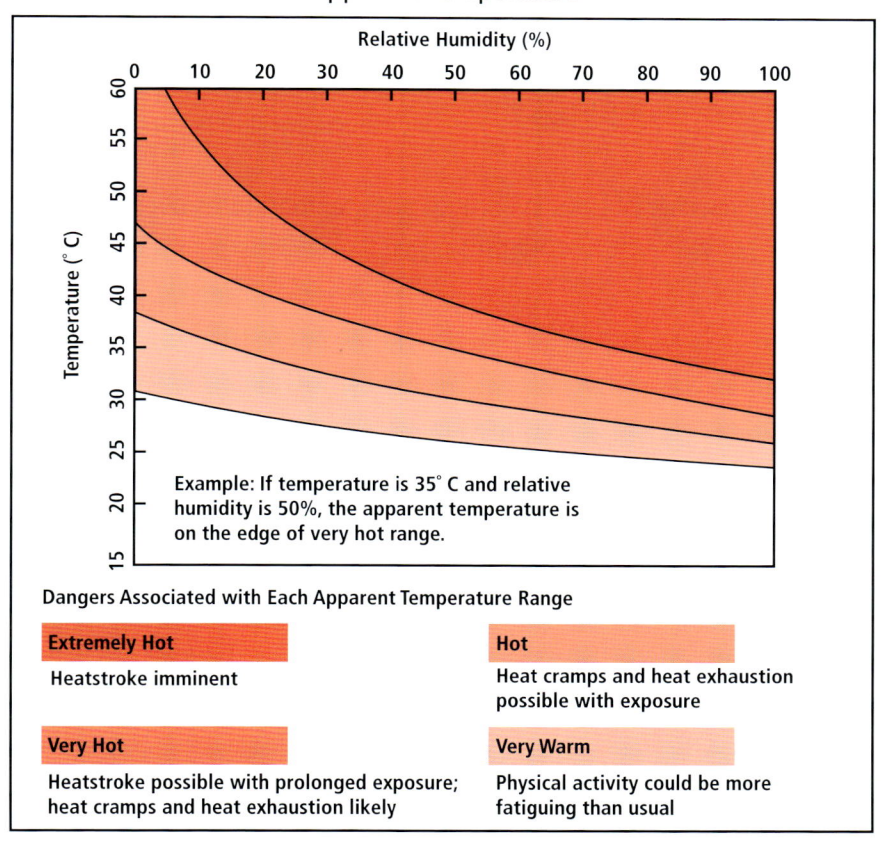

Dangers Associated with Each Apparent Temperature Range

Extremely Hot	Hot
Heatstroke imminent	Heat cramps and heat exhaustion possible with exposure

Very Hot	Very Warm
Heatstroke possible with prolonged exposure; heat cramps and heat exhaustion likely	Physical activity could be more fatiguing than usual

Fig. 7-9

Ontario's Climate

Average summer temperatures can differ by 10° to 15° C across the province. These climatic averages conceal the fact that the hottest place in the province can often be in the northern regions and not in the south, as one might expect. For example, one might be astonished to learn that Moosonee, along the shores of James Bay, reached 38° C in July 1975 during a southerly downslope flow. One would be less surprised by the fact that Toronto reached 40.6° C on three consecutive days in July 1936. At the same time, the temperature in Atikokan and nearby Fort Frances in northwestern Ontario reached a provincial record of 42.2° C. This Ontario all-time record-high temperature is shared with Biscotasing (northwest of Sudbury), where temperatures reached 42.2° C in July 1919. Somewhat surprisingly, the hottest temperatures in Ontario were all set in areas north of the Great Lakes. The heat wave of July 15–17, 1936, is remembered as the deadliest in Canadian history, with 1180 people dying in Manitoba and Ontario. The actual weather can be concealed by the climate and is where meteorologists really add value to the predictions.

Average winter temperatures show a larger range of almost 25° C. The moderating effects of the Great Lakes are clearly revealed anywhere downwind from the water. Areas south and west of Barrie are virtually surrounded by three of the Great Lakes and have average winter temperatures just below freezing.

Annual number of days in Toronto with temperatures over 30° C
1971–2000

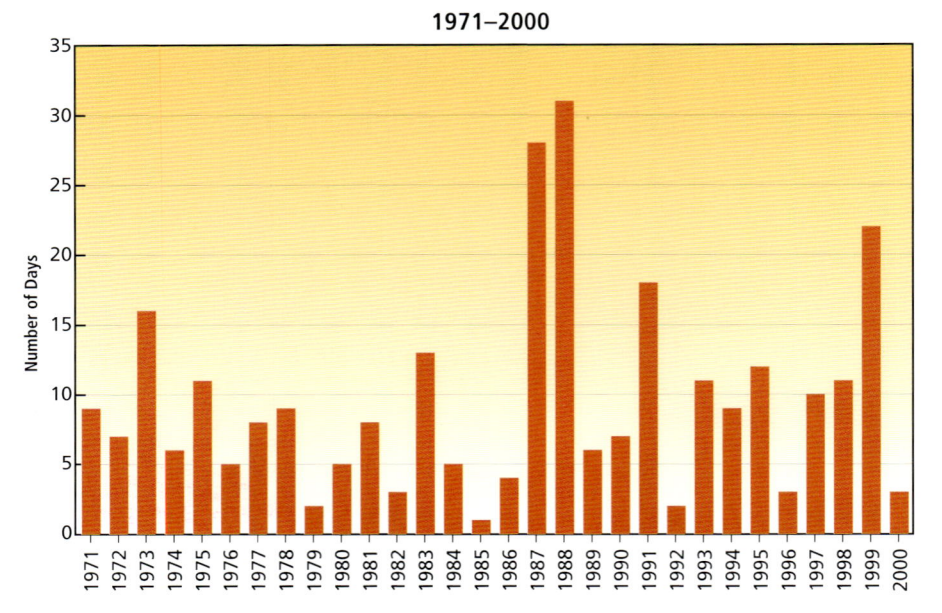

Fig. 7-10

Ontario's Climate

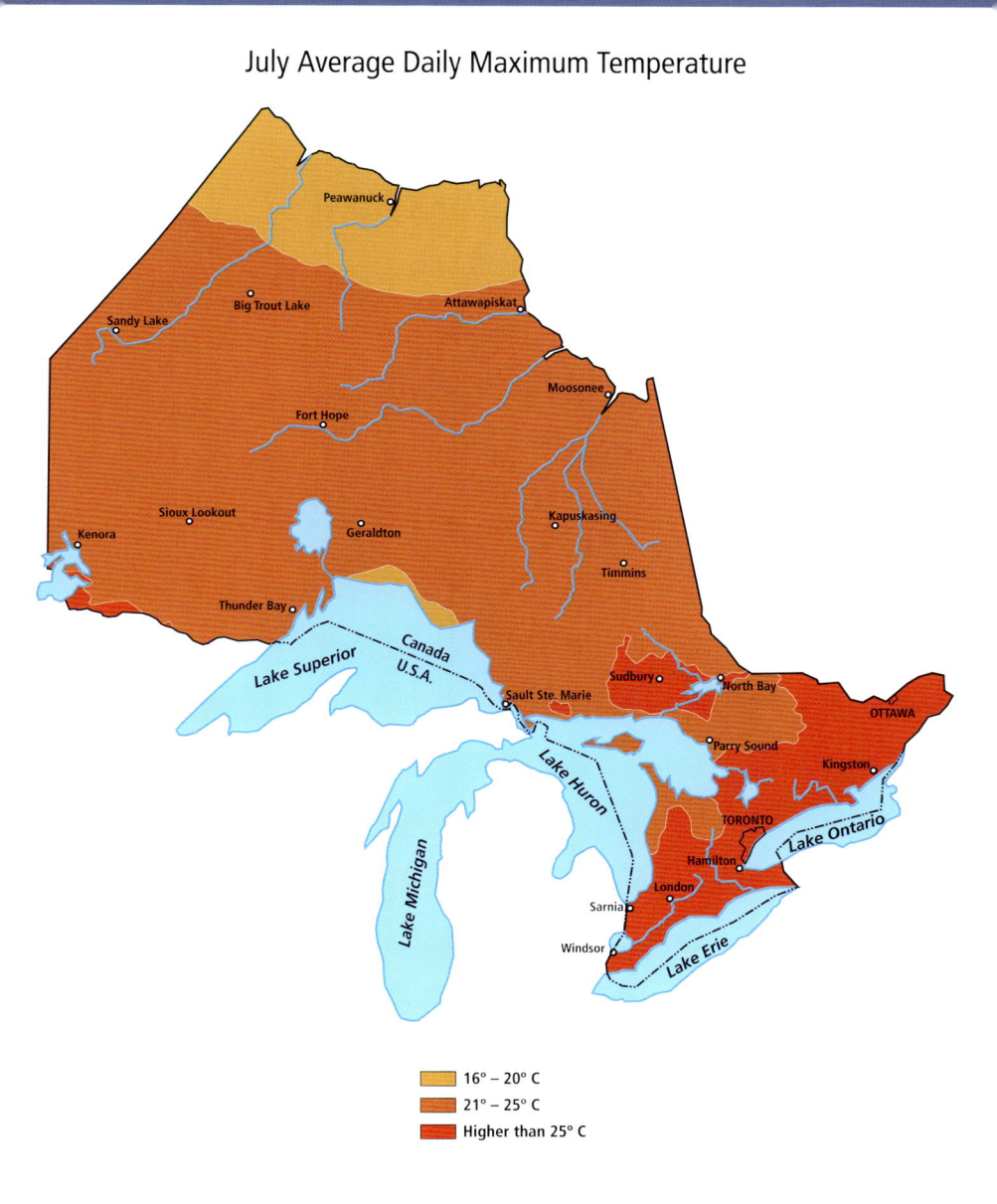

July Average Daily Maximum Temperature

- 16° – 20° C
- 21° – 25° C
- Higher than 25° C

Fig. 7-11

Ontario's Climate

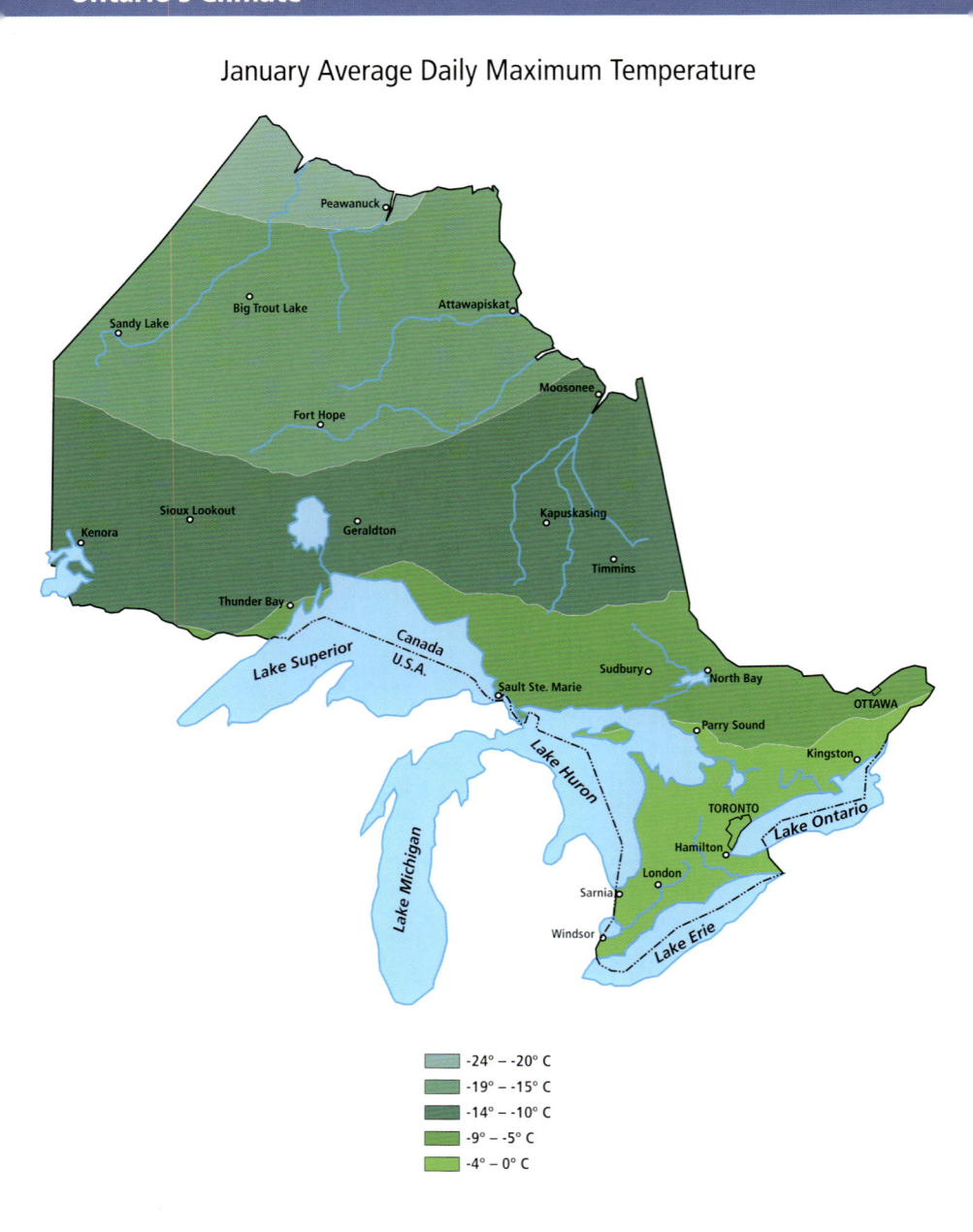

January Average Daily Maximum Temperature

- −24° – −20° C
- −19° – −15° C
- −14° – −10° C
- −9° – −5° C
- −4° – 0° C

Fig. 7-12

Ontario's Climate

Urban Heat Islands

Cities have a modifying effect and create local microclimates, known as urban heat islands. The most apparent effect is on temperature because buildings, factories and moving vehicles emit large amounts of waste heat, and concrete buildings, asphalt parking lots and roadways absorb heat during daylight hours and release it overnight. The larger the urban area, the more pronounced the heat island. The heat island effect in Ontario's largest cities amounts to several degrees Celsius and is most pronounced in winter, especially with minimum overnight temperatures.

Heating Degree Days

A practical application of temperature statistics is the calculation of the Heating Degree Day (HDD). Heating Degree Days (see Figure 7-13) are highly correlated with the amount of energy required to heat buildings to a comfortable level. Heating is generally required when external temperatures are below some reference value, usually 18° C. The Heating Degree Day is calculated by subtracting the average daily temperature from 18. For example, if the average temperature for a day were 10° C, then the HDD value would be 8; if the day's average were -10° C, then the HDD value would be 28. Daily temperatures above 18 have an HDD of zero. The HDD values can be summed over a year to give a useful estimate of heating energy requirements. As shown in Figure 7-14, Heating Degree Days range from their lowest values of near 3700 in the Windsor to Sarnia corridor to more than 8250 along the Hudson Bay. This means that it would take more than twice the amount of energy to heat a home in Fort Severn, located on the Hudson Bay (and the most northern community in Ontario), as it would in Windsor, in the extreme south of the province. Warming trends have been noted in the HDD for some cities, probably as a result of the urban heat island effect. Any patterns in HDD maps are complicated by natural variability in the climate as well as El Niño and La Niña events in the tropical Pacific Ocean.

What Are Heating Degree Days?

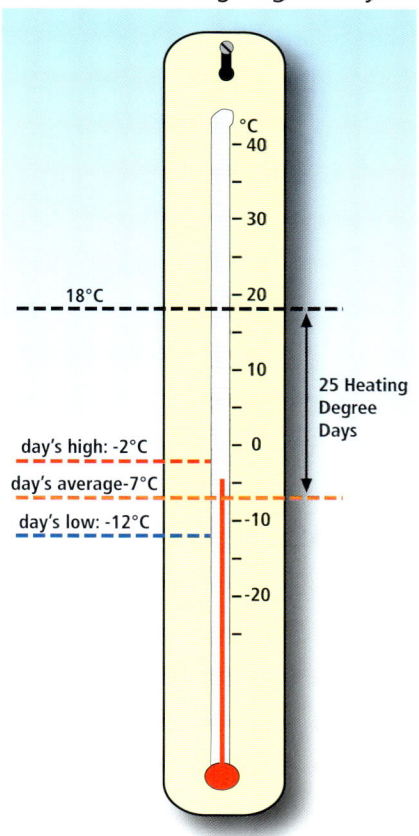

Fig. 7-13 The day's average temperature is subtracted from 18° C to get the Heating Degree Days.

Ontario's Climate

Annual Total Heating Degree Days Below 18° C

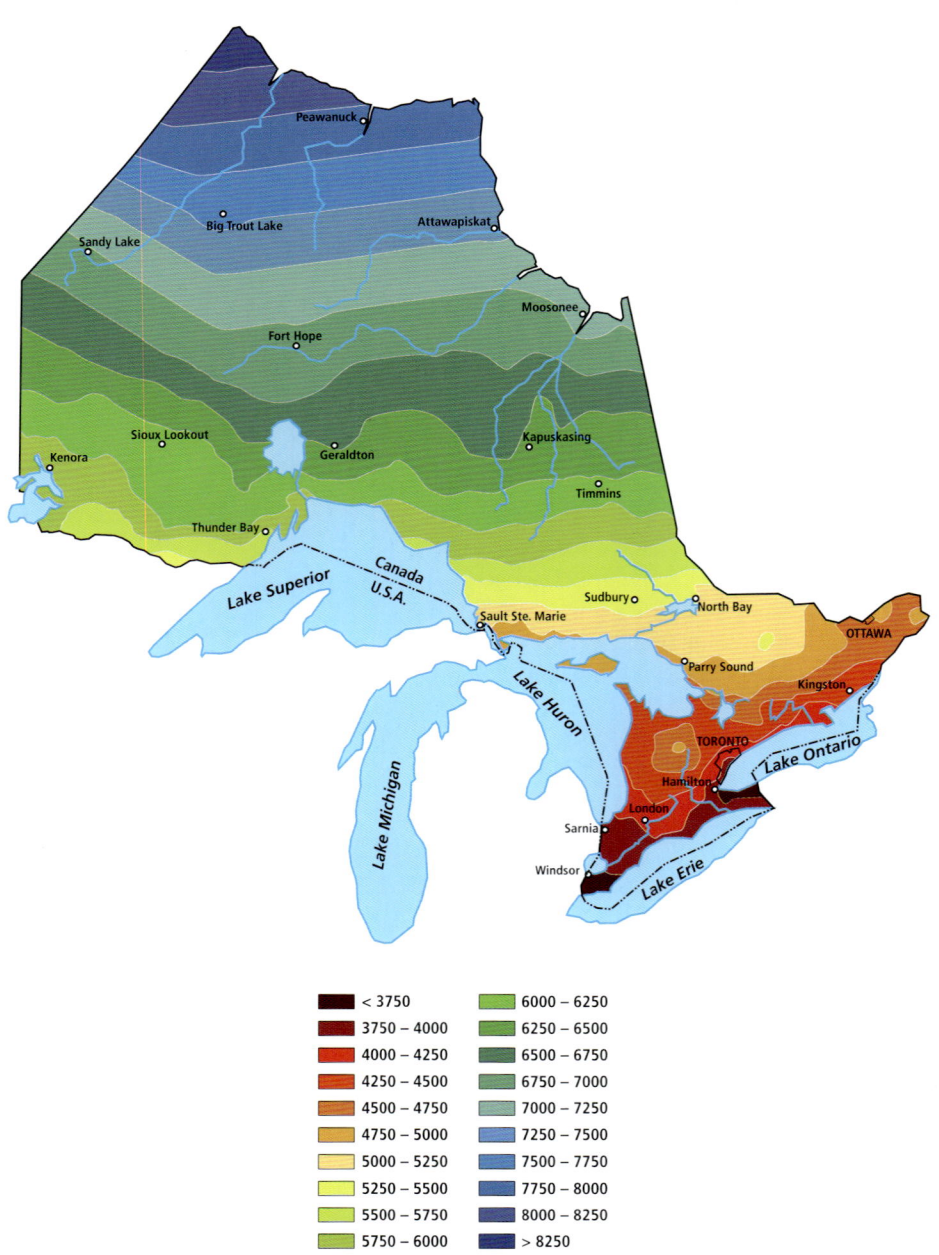

< 3750	6000 – 6250
3750 – 4000	6250 – 6500
4000 – 4250	6500 – 6750
4250 – 4500	6750 – 7000
4500 – 4750	7000 – 7250
4750 – 5000	7250 – 7500
5000 – 5250	7500 – 7750
5250 – 5500	7750 – 8000
5500 – 5750	8000 – 8250
5750 – 6000	> 8250

Fig. 7-14

Ontario's Climate

Growing Degree Days

The summer growing season varies greatly across Ontario. The Growing Degree Day (GDD) is a useful measure of the amount of heat available to initiate and sustain crop growth. The GDD is related to how far the average daily temperature varies from a specific reference temperature. The daily values are accumulated over the growing season. The reference temperature depends on the type of plant, but 5° C is often used for general plant growth. Figure 7-15 shows the Growing Degree Days above 5° C for Ontario. The area between Sarnia and Hamilton along the north shore of Lake Erie has the most Growing Degree Days. Another swath along the north shore of Lake Ontario from Lake Simcoe to Cornwall also has a high GDD. Plant growth, of course, depends on the frost-free period as well. The frost-free period is defined as the number of days between the last frost in spring and the first frost in autumn. Figure 7-16 shows the average frost-free period across the province. Much of southern Ontario has up to 200 frost-free days per year.

"Blue Sky Lake" by Phil Chadwick

Ontario's Climate

Annual Total Growing Degree Days Above 5° C

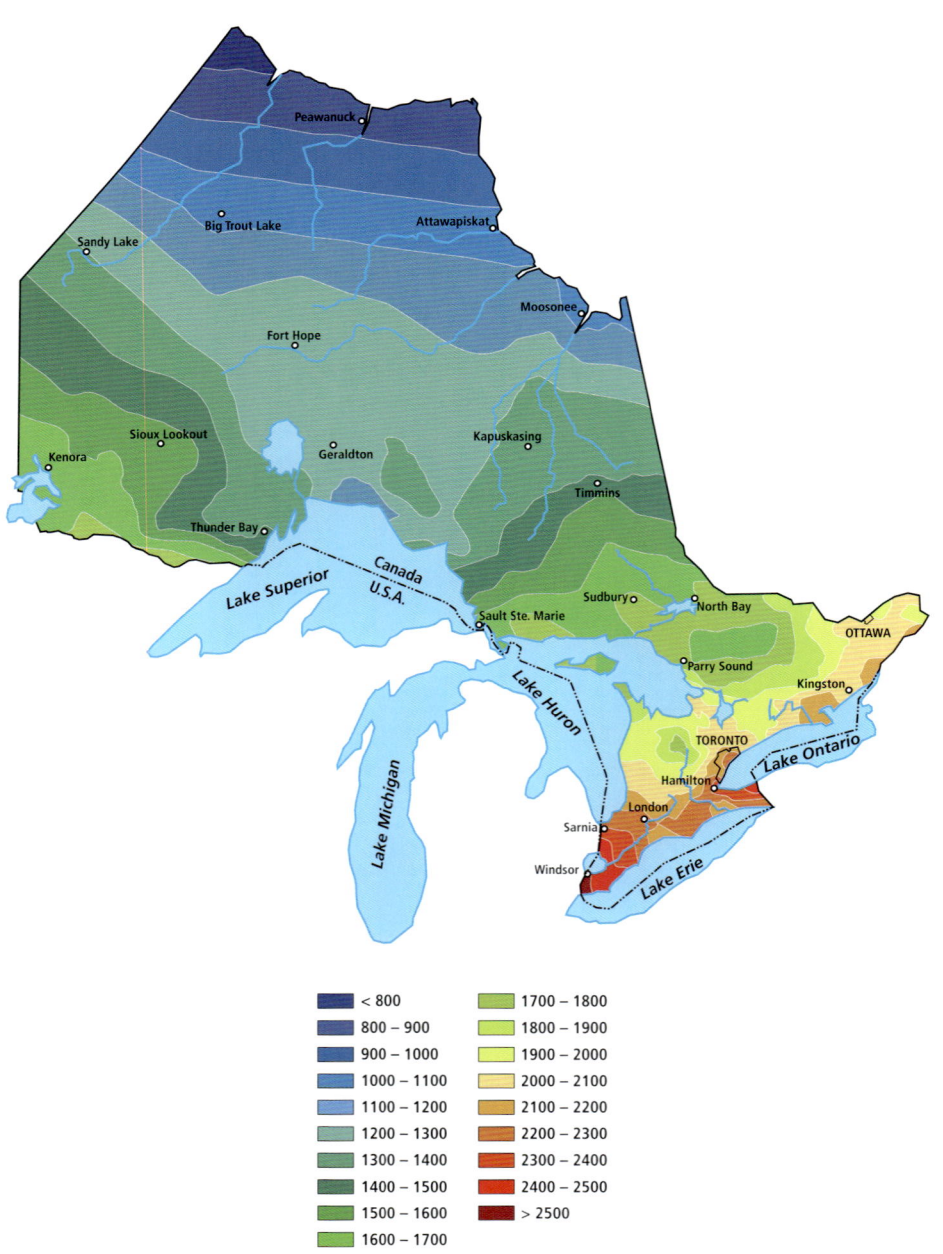

Fig. 7-15

Ontario's Climate

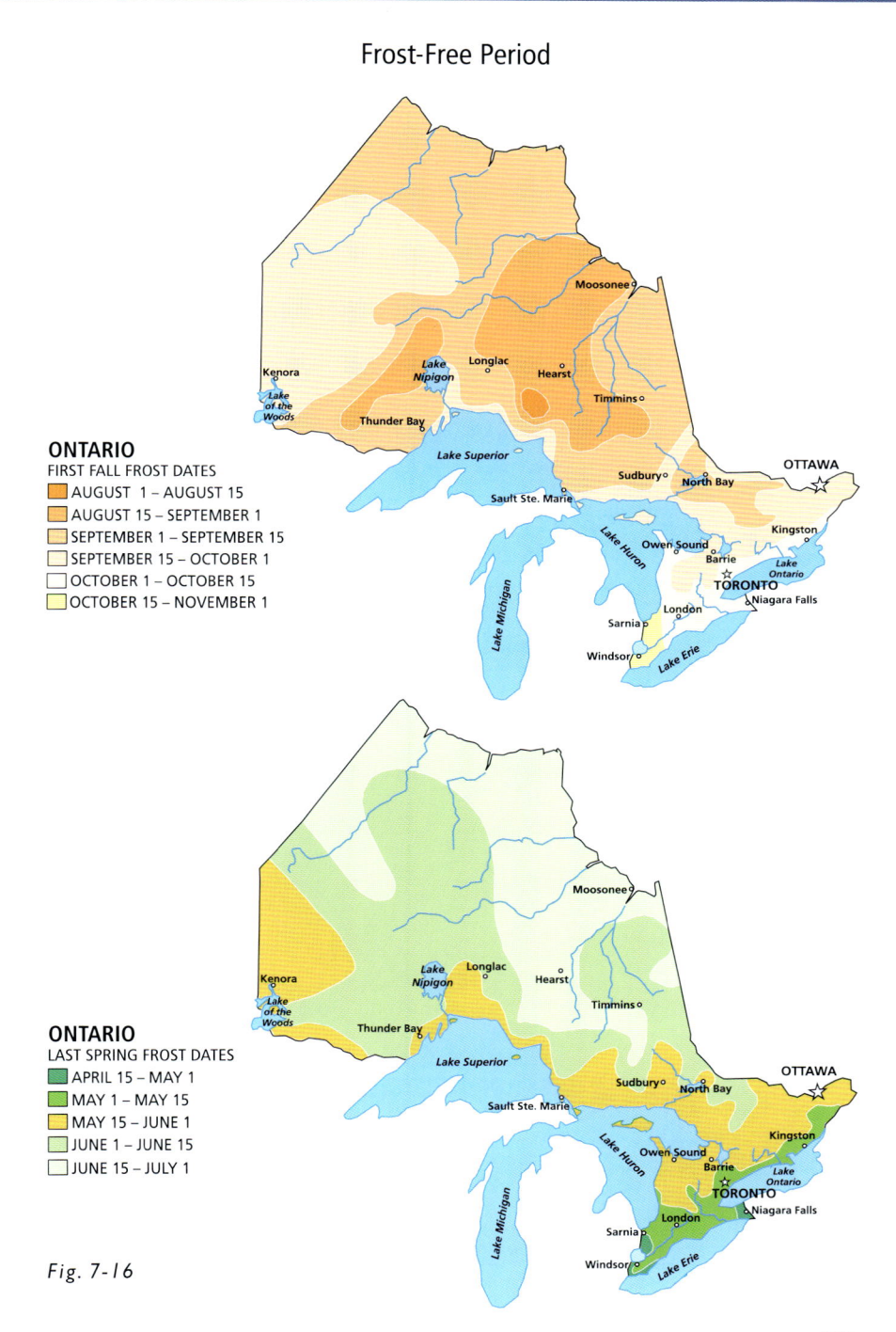

Frost-Free Period

ONTARIO
FIRST FALL FROST DATES
- AUGUST 1 – AUGUST 15
- AUGUST 15 – SEPTEMBER 1
- SEPTEMBER 1 – SEPTEMBER 15
- SEPTEMBER 15 – OCTOBER 1
- OCTOBER 1 – OCTOBER 15
- OCTOBER 15 – NOVEMBER 1

ONTARIO
LAST SPRING FROST DATES
- APRIL 15 – MAY 1
- MAY 1 – MAY 15
- MAY 15 – JUNE 1
- JUNE 1 – JUNE 15
- JUNE 15 – JULY 1

Fig. 7-16

Ontario's Climate

Combinations of Weather Elements

Human discomfort is usually caused by a combination of weather elements. In winter, the combination of cold temperatures and wind has a pronounced chilling effect on skin. In summer, high humidity combined with warm temperatures prevents our bodies' evaporative cooling mechanism from performing effectively.

Wind Chill

The wind chill index has been developed to measure this cooling effect and to signal conditions that may be harmful. The wind chill calculation table provides "feels like" values in degrees Celsius, indicating how cold the temperature feels once the strength of the wind has been factored in. Wind chill values can fall below -30° C at all locations in Ontario during winter.

> The coldest wind chill recorded in Ontario was in Thunder Bay on January 10, 1982. The temperature was -36° C, but 54 km/h winds produced a wind chill of -58° C.

Wind Chill Calculation Table

T air / V_{10}	5	0	-5	-10	-15	-20	-25	-30	-35	-40	-45	-50
5	4	-2	-7	-13	-19	-24	-30	-36	-41	-47	-53	-58
10	3	-3	-9	-15	-21	-27	-33	-39	-45	-51	-57	-63
15	2	-4	-11	-17	-23	-29	-35	-41	-48	-54	-60	-66
20	1	-5	-12	-18	-24	-30	-37	-43	-49	-56	-62	-68
25	1	-6	-12	-19	-25	-32	-38	-44	-51	-57	-64	-70
30	0	-6	-13	-20	-26	-33	-39	-46	-52	-59	-65	-72
35	0	-7	-14	-20	-27	-33	-40	-47	-53	-60	-66	-73
40	-1	-7	-14	-21	-27	-34	-41	-48	-54	-61	-68	-74
45	-1	-8	-15	-21	-28	-35	-42	-48	-55	-62	-69	-75
50	-1	-8	-15	-22	-29	-35	-42	-49	-56	-63	-69	-76
55	-2	-8	-15	-22	-29	-36	-43	-50	-57	-63	-70	-77
60	-2	-9	-16	-23	-30	-36	-43	-50	-57	-64	-71	-78
65	-2	-9	-16	-23	-30	-37	-44	-51	-58	-65	-72	-79
70	-2	-9	-16	-23	-30	-37	-44	-51	-58	-65	-72	-80
75	-3	-10	-17	-24	-31	-38	-45	-52	-59	-66	-73	-80
80	-3	-10	-17	-24	-31	-38	-45	-52	-60	-67	-74	-81

T air = air temperature in ° C and V_{10} = observed wind speed at 10 m elevation, in km/h.

Frostbite Guide

Low risk of frostbite for most people
Increasing risk of frostbite for most people in 10 to 30 minutes of exposure
High risk for most people in 5 to 10 minutes of exposure
High risk for most people in 2 to 5 minutes of exposure
High risk for most people in 2 minutes of exposure or less

Fig. 7-17

Ontario's Climate

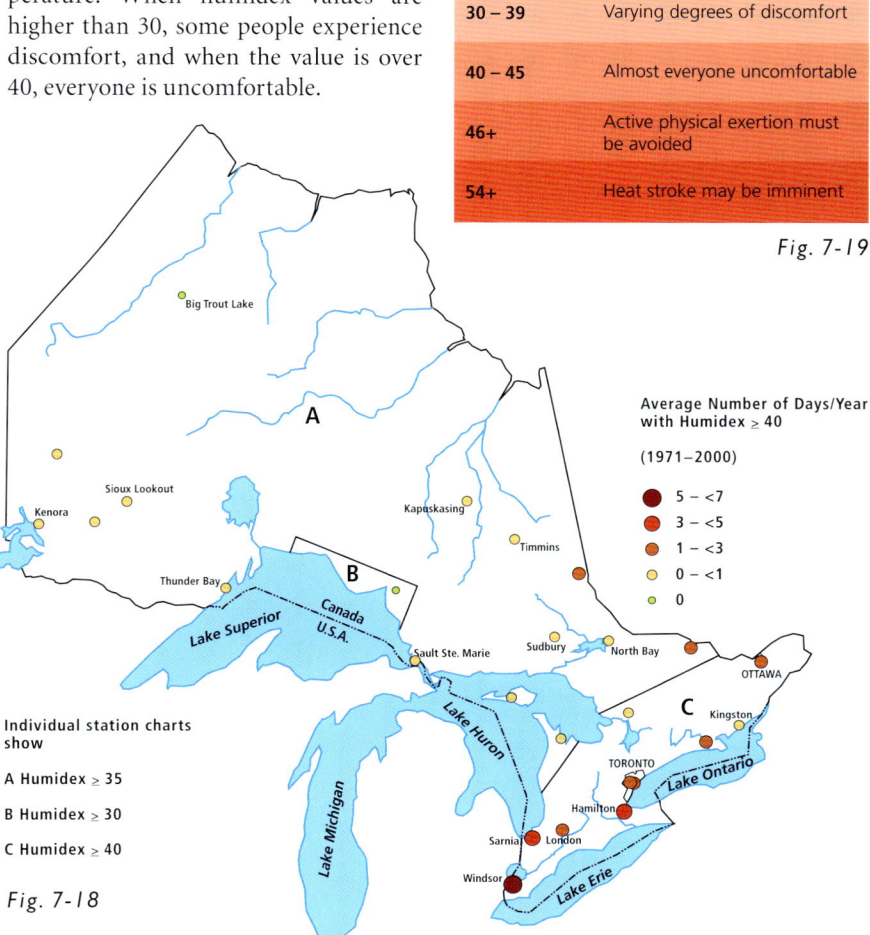

Humidex

The humidex is a Canadian innovation, first used in 1965. The degree of discomfort humans experience in warm weather depends on the rate at which our bodies can lose heat through the evaporation of perspiration. This natural cooling mechanism is compromised when atmospheric humidity is high enough to inhibit evaporation. The humidex index (Figure 7-19) combines air temperature and relative humidity into a number that is a "feels like" temperature. When humidex values are higher than 30, some people experience discomfort, and when the value is over 40, everyone is uncomfortable.

Extremely high readings are rare except in the southern regions of Ontario, Manitoba and Québec. Generally, the humidex decreases toward the north. Of all Canadian cities, Windsor, Ontario, has had the highest recorded humidex measurement: 52.1 on June 20, 1953. On average, humidex values exceed 40 on 6 days per year in Windsor.

HUMIDEX (°C)	DEGREE OF DISCOMFORT
20 – 29	Comfortable
30 – 39	Varying degrees of discomfort
40 – 45	Almost everyone uncomfortable
46+	Active physical exertion must be avoided
54+	Heat stroke may be imminent

Fig. 7-19

Individual station charts show

A Humidex ≥ 35
B Humidex ≥ 30
C Humidex ≥ 40

Fig. 7-18

Ontario's Climate

Precipitation

The continental nature of the climate is clear in northern Ontario, where most precipitation falls during the warm season. Roughly 65 percent of the annual precipitation for northern Ontario falls in summer. The modified continental climate near the Great Lakes shows a much more even annual distribution of precipitation. There is no clear seasonal variation in precipitation across southern Ontario. Overall, the southern portion of the province tends to get more precipitation than the north.

Cold Lows

Rainfall can occasionally come in the form of a prolonged drenching that is usually associated with what is known as a cold low. This type of circulation system is most common in spring and fall over Ontario and can remain almost stationary for several days over the lower Great Lakes. Winds flowing counterclockwise around a cold low often contain an abundance of moisture, and rainfall is persistent. The highest recorded one-day rainfall—the Harrow storm of July 19–20, 1989—produced a 24-hour rainfall record of 264.2 millimetres, surpassing any official 24-hour rainfall previously recorded east of Vancouver Island and exceeding even Hurricane Hazel's highest 48-hour total by 40 percent. The Harrow rainfall was associated with a cold low circulation. This type of storm can really soak the surface soil layer, leading to flooding in the river basins of southwestern Ontario.

Snow

Ontario ranks about midway in the provincial pack for snowfall. On average, Ontario gets neither too much nor too little snow—the amount is just right. The big snowstorms are the result of Colorado lows that bring abundant moisture from the Gulf of Mexico. This moisture is converted into precipitation, and regions north of the warm front will get this precipitation in the form of snow.

The average annual snowfall map (Fig. 7-22) reveals a broad swath from north of Lake Superior southeast along the Ottawa Valley that receives up to 300 centimetres of snow annually. South of this swath, the precipitation of winter storms typically changes phase to freezing rain or rain. The warm air mass required to supply the moisture to create the snow doesn't normally make it north of this swath, so the snowfall tapers off toward Hudson Bay.

> **Sunshine is delicious, rain is refreshing, wind braces us up, snow is exhilarating; there is really no such thing as bad weather, only different kinds of good weather.**
>
> —John Rushkin

Ontario's Climate

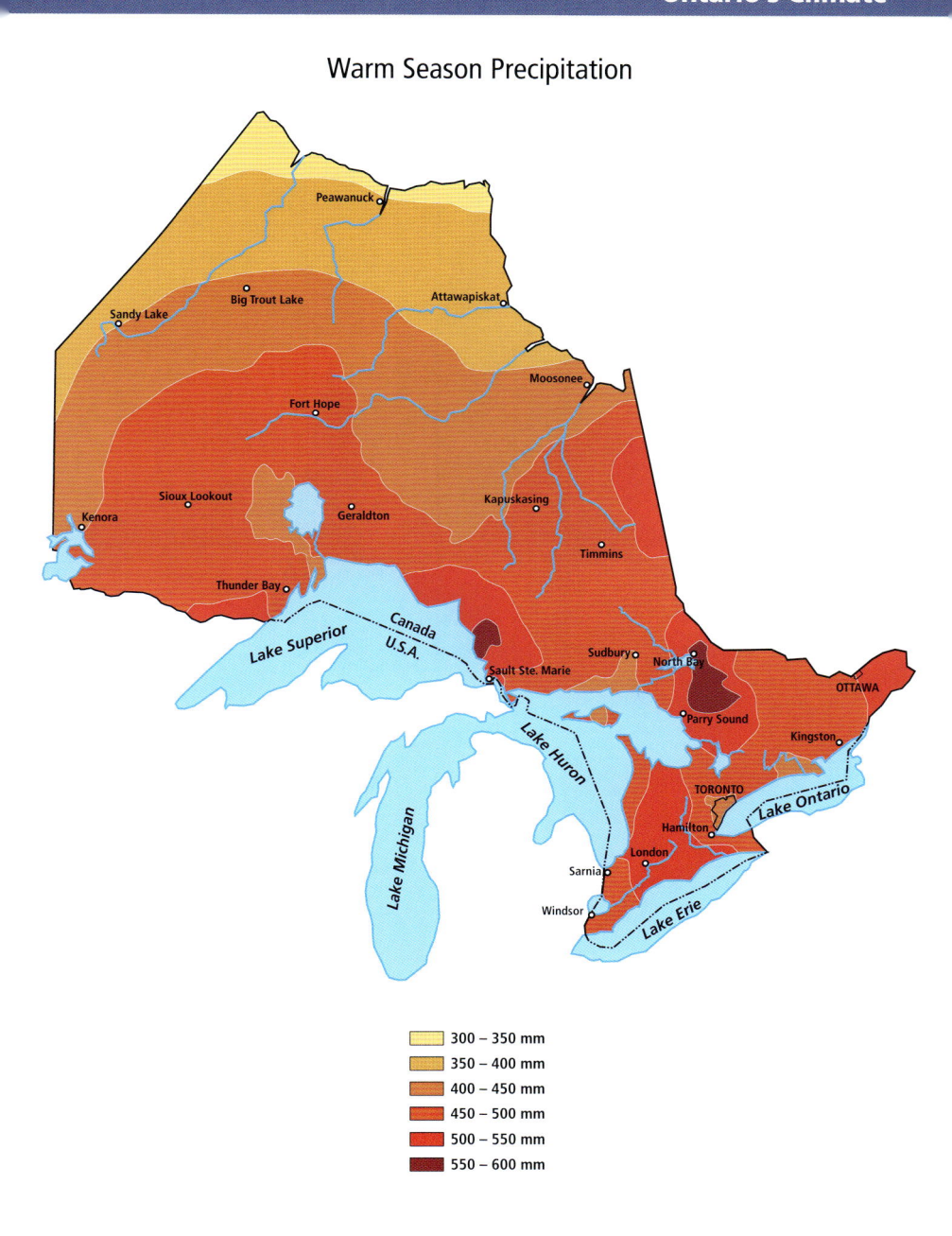

Warm Season Precipitation

- 300 – 350 mm
- 350 – 400 mm
- 400 – 450 mm
- 450 – 500 mm
- 500 – 550 mm
- 550 – 600 mm

Fig. 7-20

Ontario's Climate

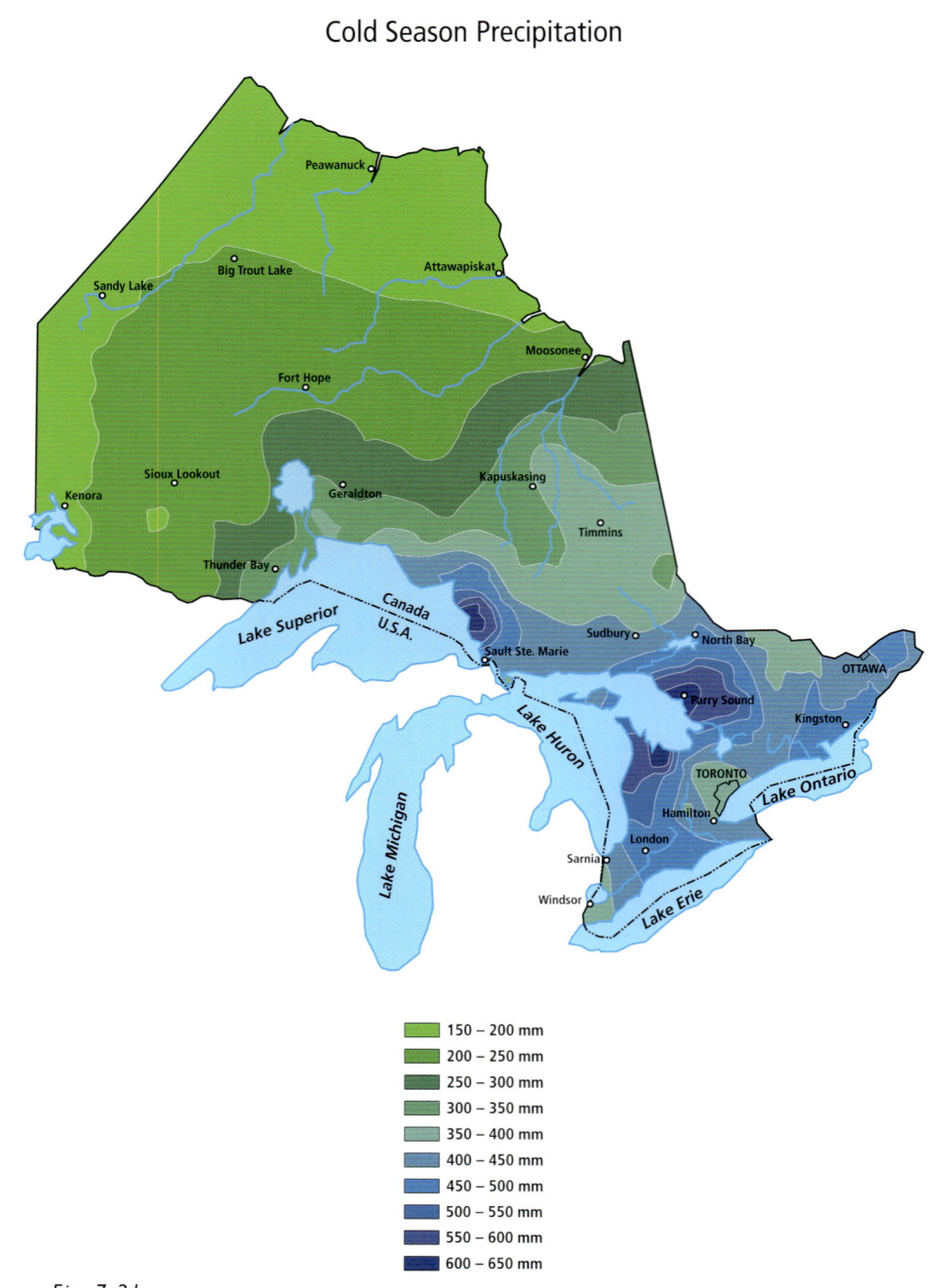

Cold Season Precipitation

■	150 – 200 mm
■	200 – 250 mm
■	250 – 300 mm
■	300 – 350 mm
■	350 – 400 mm
■	400 – 450 mm
■	450 – 500 mm
■	500 – 550 mm
■	550 – 600 mm
■	600 – 650 mm

Fig. 7-21

Ontario's Climate

Average Annual Snowfall

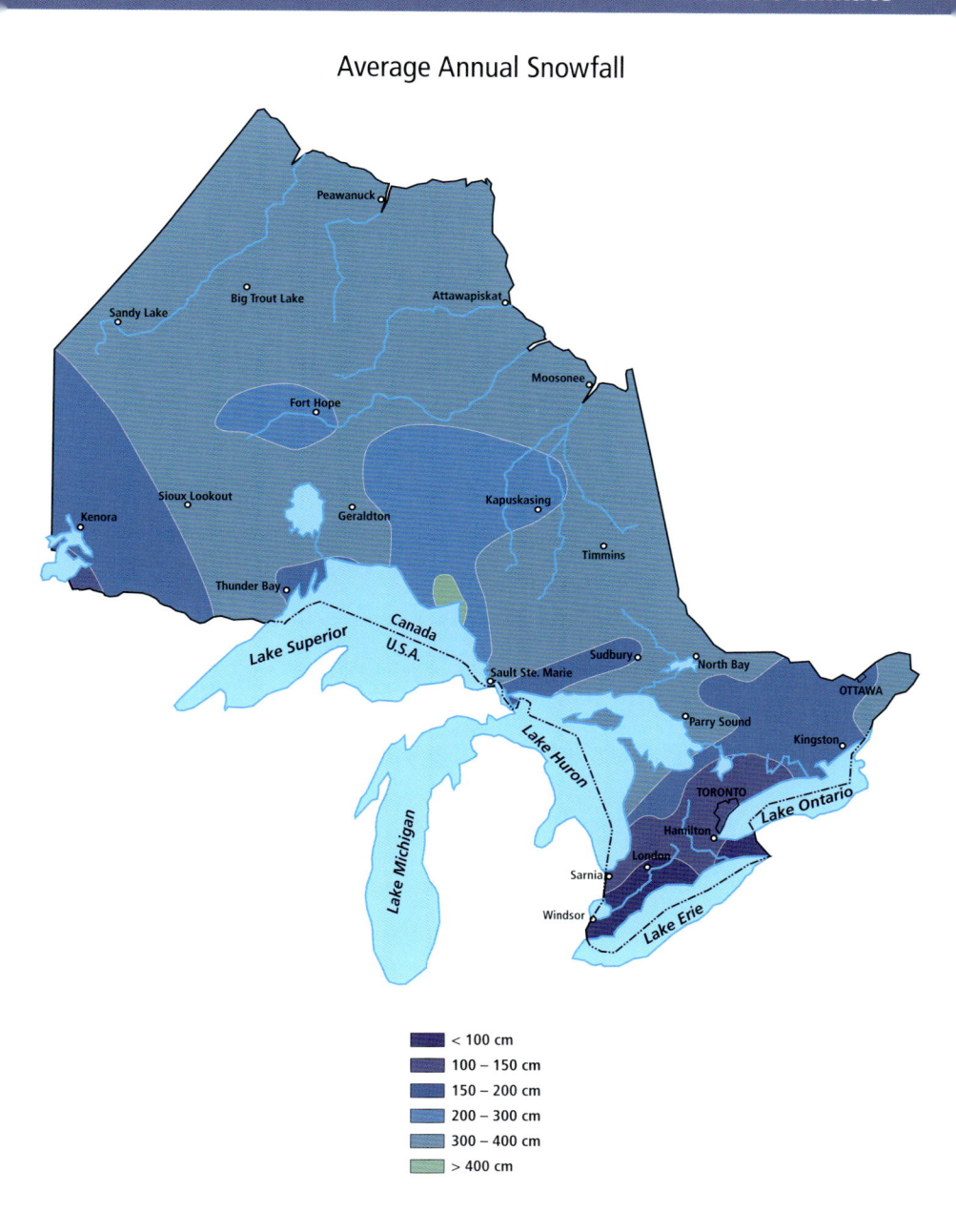

- < 100 cm
- 100 – 150 cm
- 150 – 200 cm
- 200 – 300 cm
- 300 – 400 cm
- > 400 cm

Fig. 7-22

137

Ontario's Climate

Drought

The absence of precipitation is extremely problematic, and it affects all sectors of the economy. Reduced crop yields and crop failure are obvious impacts in the agricultural sector, but soil loss through wind erosion, reduction in municipal water supply, the increased susceptibility of timber to fire and pest infestation, and the loss of wildlife habitat are also serious concerns. The droughts of 1988, mid-1997 through 1999 and the summers of 2001, 2002 and 2005 all had substantial impacts.

Recurrent extreme drought conditions in Ontario since the 1980s have raised awareness of the importance of water resources and have stimulated drought research. Drought conditions occur because of a number of variables, including the lack of precipitation, the variability of precipitation with time (i.e., precipitation at the wrong time), concurrent warm spells and even changes in the ability of society and wildlife to adapt to a changing environment.

There are several methods and indicators to measure drought. The Palmer Drought Index (PDI) was used extensively in western Canada to depict drought conditions. The empirical PDI has some limitations because it was designed for use in the United States and does not consider variability in precipitation, which is a characteristic of Ontario precipitation.

The Ontario Low Water Response (OLWR) plan is based on several indicators (nine or more) and threshold criteria. When OLWR criteria relating to the

Record Wet and Record Dry in Ontario during 2007

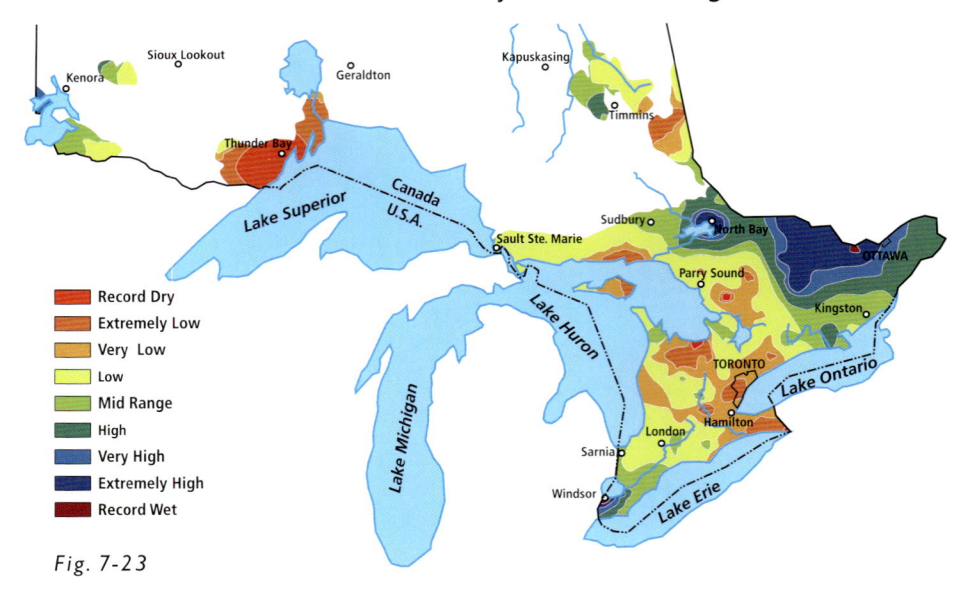

Fig. 7-23

Ontario's Climate

severity of low water and precipitation deficit conditions are met, specific response actions are put into place. Over the larger geographical study area of Ontario, regional patterns of low precipitation are evident, with the far north receiving less annual precipitation but historically also appearing to be less prone to OLWR-indicated drought conditions. Although northern Ontario receives less precipitation, the precipitation tends to be spread out over more days (more consecutive wet days and fewer consecutive dry days). In southern Ontario, the situation is reversed, with higher annual precipitation totals, along with a much higher frequency of OLWR-indicated drought conditions. The consecutive-day analysis demonstrates that southern Ontario experiences a greater number of consecutive dry days than do the northern regions of the province. The OLWR-indicated drought conditions are typically experienced over the southern and extreme northwestern regions of Ontario. Seasonally, the OLWR-indicated drought conditions are most likely over southwestern Ontario during summer and autumn while they are more of an autumn and winter concern over northwestern Ontario.

The North American Drought Monitor (NA-DM) became operational in 2003 to recognize the continental scope and importance of drought. Like the OLWR, it is based on a combination of drought indices. It demonstrates the vital importance of continental water resources and the seriousness of a lack of water.

Precipitation Thresholds for OLWR

Level I	Level II	Level III
18-month precipitation <80% of average precipitation	18-month precipitation <60% of average precipitation	18-month precipitation <40% of average precipitation
or	or	or
3-month precipitation <80% of average precipitation	3-month precipitation <60% of average precipitation	3-month precipitation <40% of average precipitation
	or	or
	1-month precipitation <60% of average precipitation	1-month precipitation <40% of average precipitation
	or	
	<7.6 mm rain for 2 weeks in high demand areas or 3 weeks in moderate demand areas	

Fig. 7-24

Ontario's Climate

OLWR-level Toronto Drought

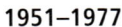

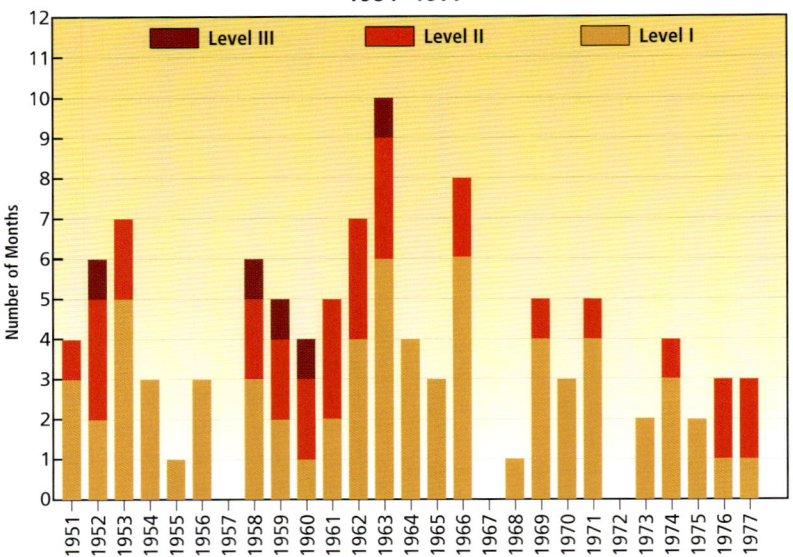

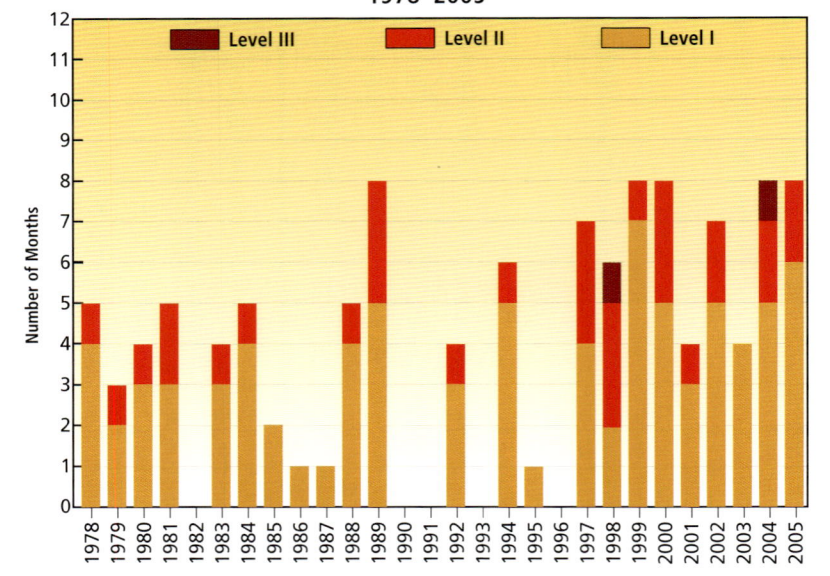

Fig. 7-25

Ontario's Climate

Sunshine

Toronto experiences an average of 2038 sunshine hours, or 44 percent of possible hours, most of it during the warmer weather season.

In the three-month stretch from September to November 2006, Ontario recorded its gloomiest autumn in 29 years, with a total of just 343 hours of sunlight instead of the average 475 hours. Toronto began recording the amount of sunshine back in 1957. The gloomy fall of 2007 was surpassed only by that of 1977, which had the second-gloomiest autumn with a dismal 323 hours, and 1970, which had the gloomiest autumn ever with 312 hours.

> On average, Thunder Bay enjoys the sunniest winters in the province.

Average Annual Hours of Sunshine

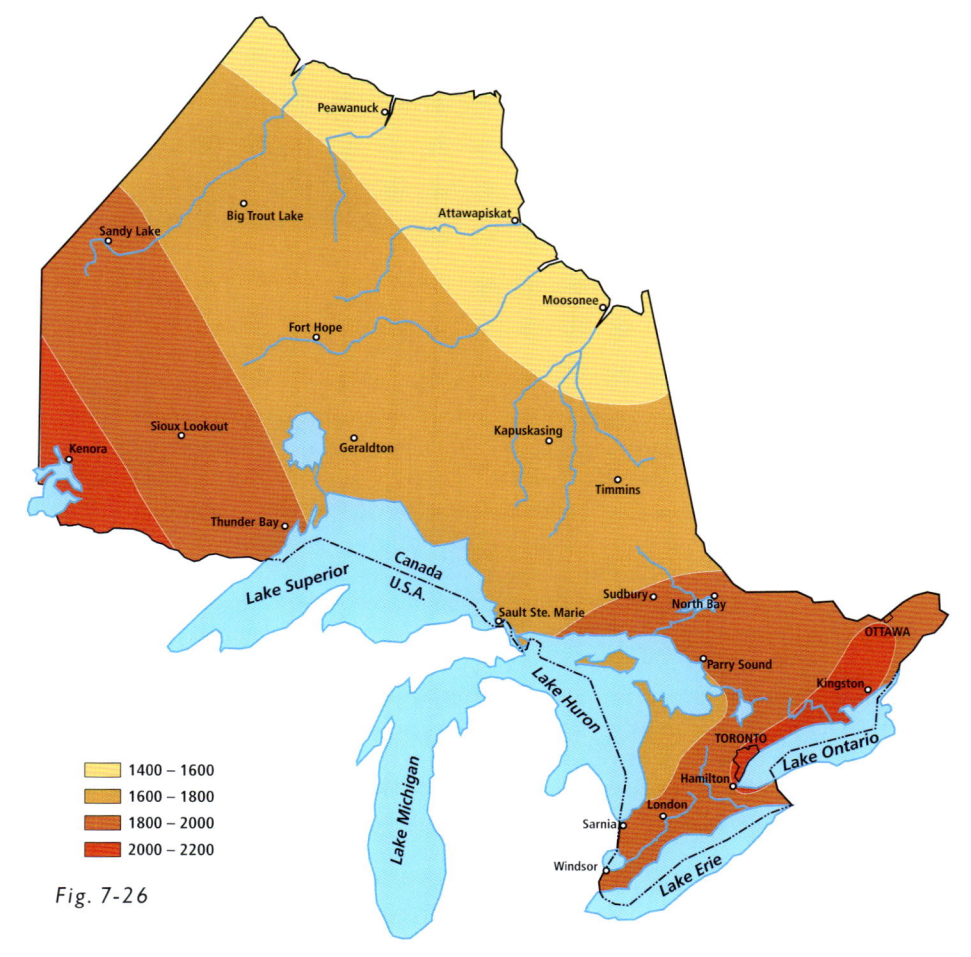

- 1400 – 1600
- 1600 – 1800
- 1800 – 2000
- 2000 – 2200

Fig. 7-26

Chapter 8: Storms

The Storm of the Century '93, Ice Storm '98 and Hurricane Hazel '54 all made the top 10 of the most severe eastern North American weather events since 1948 (in positions 3, 5 and 9, respectively). And all affected Ontario. Ontarians are clearly subjected to all categories of weather—severe and otherwise. The storm characteristics are determined by their warm inflow (heat and moisture energy) and their location relative to the jet stream (wind energy). The jet stream is always moving with the long wave pattern, but preferential locations and orientations can be linked to ENSO (see Chapter 7), which is why I have included this information with each of the historic storms noted.

Summer storms tend to move rather slowly, averaging only 20 kilometres per hour. Even the winds in the jet stream are slow (typically only about 150 kilometres per hour). The low pressure area that characterizes the storm also defines the convection that develops. Severe summer convection is common across the entire province but gets reported most in the south (due to larger population and infrastructure densities). Tornadoes, hail, downburst winds and torrential rain are common. In a typical summer, upward of 200 severe convective events will be observed in Ontario. Many more severe events go undetected, occurring away from urban areas and within the vast forests of the province. It is estimated

Storms

that only 10 percent or less of the severe convective events that occur are actually observed. Ontario also has a few tornado alleys, with the most important occupying a broad swath from Windsor northeast to Lake Simcoe, remaining north of Highway 401.

At 50 kilometres per hour or more, winter storms typically move fast. The jet stream winds are even faster (up to 300 kilometres per hour or more) and supply the storms with lots of wind energy. Winter severe weather is everywhere, though the blizzards and extreme cold are typically confined to the northern half of the province. Freezing rain is common along and north of the Oak Ridges Moraine to the valleys of the St. Lawrence and Ottawa rivers. Heavy snow is most likely over central Ontario and "Cottage Country," because farther south the precipitation tends to change to freezing rain or rain, meaning there is less snow to accumulate.

Both summer and winter storms move slower as they increase in size and change their energy relationship, going from being tied to faster-moving short waves to being associated with the slower-moving longer wavelength and deeper troughs.

Wind storms off the Great Lakes and the remains of hurricanes are big weather producers in autumn. Otherwise, autumn is a glorious time, with excellent weather and the warm waters of the Great Lakes perpetuating the feel of summer.

Spring is another transitional period, but this time the cold Great Lakes prolong the chilly, damp feel of winter along the lakeshore. Often, just as we start to think of the coming warm weather and the pleasures of gardening, boating or returning to the golf course, Old Man Winter throws us a nasty surprise—the

Typical Ontario Winter Storm

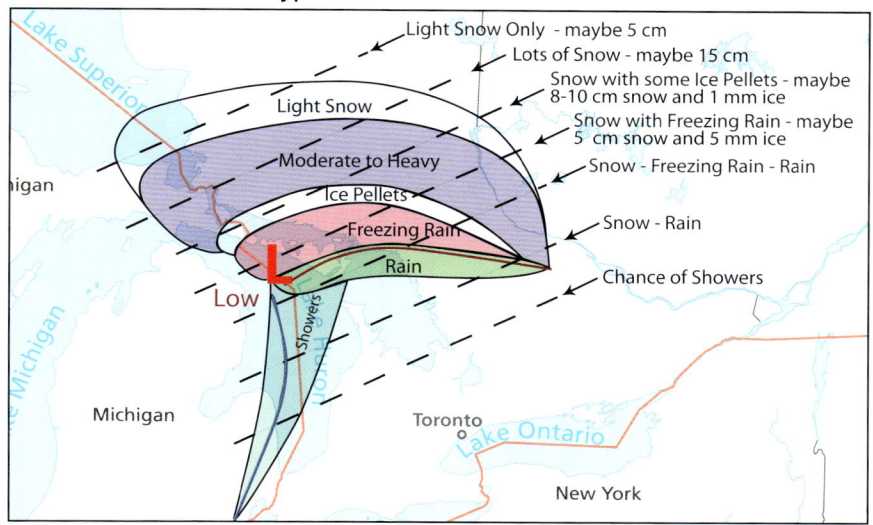

Fig. 8-1

Storms

late-spring snowstorm. David Phillips, Canada's most well-known climatologist, coined the term "spring whitewash" as an appropriate description. These storms do not happen every year in Ontario, but snow sometimes falls as late as Mother's Day.

Weather is still given top billing in the media. Everyone discusses the weather with varying degrees of authority. The communication of meteorological information remains a challenge. There is a large and growing gap between what meteorologists know and the weather information that is communicated. This information gap—which cannot be overcome by more misleading and uninformative weather icons—has a huge impact on the safety and security of Ontarians. Ontario sits in the middle of a battle of the air masses, and meteorologists follow and anticipate the daily skirmishes with precision and accuracy. But it is difficult to get the information out and even more challenging to have people take action based on accurate predictions. My own brother discounted my warnings of the Ice Storm of '98, which I had issued days ahead of the event. He didn't buy a backup generator and spent 14 very long and very cold nights in the dark without power and water. He had plenty of ice for cold drinks, though.

The '93 Superstorm really was the "storm of the century." Other weather systems were given this title, but looking back, 1993 really earned the dubious distinction.

Winter Storms
Ice Storms

In Ontario, ice storms are arguably the most severe winter weather event. Significant freezing rain events occur a dozen times each winter across Ontario, but even a brief period of freezing rain can be treacherous. On average, Ottawa receives freezing precipitation on 12 to 17 days a year, with the freezing rain generally lasting only a few hours. The nation's capital can expect 45 to 65 hours of freezing rain annually.

Ice Storm '98 (an El Niño year) began when a low pressure warm front from Texas and a high pressure arctic cold front converged simultaneously over southern Ontario. When the air masses collided, the warm air rose while the frigid air clung to the earth's surface. Snow melted at mid-level, and the raindrops became supercooled in the low level arctic air—primed to freeze on contact at ground level. The arctic front remained virtually stationary for four days while three distinct waves of moisture surged aloft along the boundary between the air masses. Overcast skies continued between significant freezing rain events, which were linked by periods of freezing drizzle and snow grains. With no sun and a fresh supply of arctic air, the delicate balance of warm and cold required for freezing rain persisted, and the ice accumulated. On one weather map, I analyzed freezing rain in a continuous band from Illinois to Nova Scotia—simply amazing!

The severity of ice storms can be determined by the ice accumulation, the duration, the location and the extent of the

Storms

area affected. These criteria make Ice Storm '98 the worst to hit Canada in recent memory. The deviation of Ice Storm '98 from the "normal" winter weather places it as number 3 in the objective list of most severe storms to affect eastern North America since 1948. The total water equivalent of precipitation that fell during the three- to four-day event exceeded 73 millimetres in Kingston, 85 millimetres in Ottawa, 108 millimetres in Cornwall and 100 millimetres in Montreal. Most of this precipitation fell as freezing rain, with some ice pellets and a bit of snow. Ice accretions from 1998 were double those of previous major ice storms in the region. Ice storms in Montreal in February 1961 and in Ottawa in December 1986 brought between 30 and 40 millimetres of ice. The impacts of Ice Storm '98 could fill another book with superlatives. The economic impact of the ice storm was estimated at over $1 billion.

Freezing Rain Accumulation January 4–10, 1998

- 40 mm
- 60 mm
- 80 mm

Fig. 8-2

145

Storms

Snowstorms

The prediction of snow accumulation is one of the most challenging tasks in weather forecasting, given the strong impact of the type of ice crystal on snow depth. The meteorologist must get not only the amount of moisture right but also the temperature and mechanism that turns it into snow—all of which change dramatically in time and space.

Generally, little (needles, hollow columns or prisms) snowflakes result in large accumulations; big (plate, branched or dendritic) snowflakes result in small accumulations. A handy little ditty sums this pattern up nicely: little snowflakes = big snow (accumulation); big snowflakes = little snow (accumulation). There are exceptions to every weather poem, but this one highlights the difference in snow

Storms

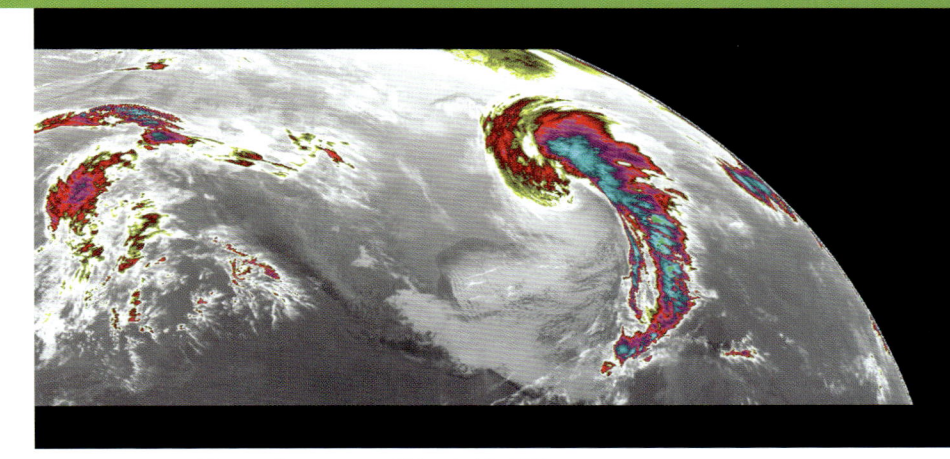

Colour-enhanced IR satellite image of March 13, 1993 superstorm

accumulation between large, slower moving synoptic-scale winter storms and convective flurries producing large dendritic flakes but that come and go quickly.

The biggest snowstorms are associated with the Colorado lows that develop in the lee of the Rockies. These lows track northeastward, taking the moisture from the warm conveyor belt off the Gulf of Mexico and converting it to snow that is dropped in a swath along and to the left of the track of the low. The Texas low variety of these storms has a more direct supply of moisture and tends to take a more southerly track over the lower Great Lakes. These storms are probable major weather producers in La Niña years and explain the extremely snowy winter of 2007–08.

These storms can also bring periods of freezing rain and heavy rain as well as strong winds. Meteorologists will typically qualify their predictions accordingly because of potential slight variations in the track of the low, which can have a huge impact on the type and amount of precipitation. The longest periods of freezing rain result when an arctic high pressure area is waiting over Québec in the path of a Colorado or Texas low.

The Storm of the Century, also known as the '93 Superstorm, No-name Hurricane, the White Hurricane or the (Great) Blizzard of 1993, was justifiably called unprecedented and historic. It was a large cyclonic storm that occurred on March 12–15, 1993, on the east coast of North America. It is unique for its intensity, massive size and wide-reaching effects. At its peak, the storm stretched from Canada to Central America. Areas as far south as central Alabama and Georgia received 10 to 15 centimetres of snow; Birmingham, Alabama, received up to 30 centimetres with isolated reports of 40 centimetres. Even the Florida Panhandle reported up to 5 centimetres of snow, with hurricane-force wind gusts and record low barometric pressures. Between Florida and Cuba, hurricane-force winds produced extreme storm surges in the Gulf of Mexico that, along with several tornadoes, killed dozens of people.

Storms

Snow, drifting and blowing snow at 4905 Dufferin Street, the location of the national headquarters of the Meteorological Service of Canada (looking west from the front steps)

Blizzards

Blizzards are relatively rare in Ontario if the definition is strictly applied. Although they can occur anywhere in the province, northern Ontario is prone to official blizzards because of the exposed landscape and colder temperatures. To meet the official definition of a blizzard, a storm must have temperatures lower than 0° C, winds stronger than 40 kilometres per hour and visibility of less than 1 kilometre because of falling and blowing

Storms

Average Annual Number of Days with Blowing Snow

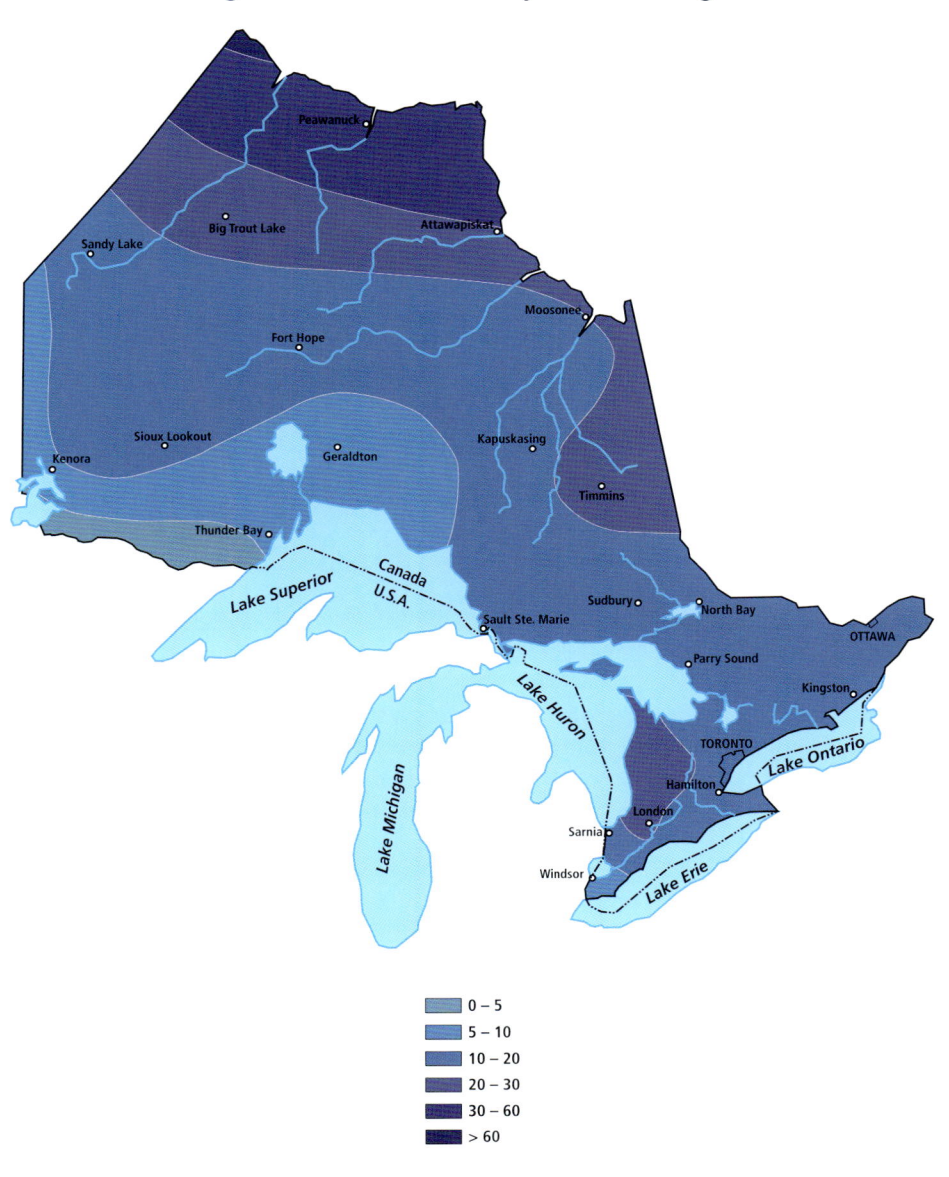

Legend:
- 0 – 5
- 5 – 10
- 10 – 20
- 20 – 30
- 30 – 60
- > 60

Fig. 8-3

Storms

Snowstorm on the crest of the Oak Ridges Moraine. Transportation stopped.

snow, and it must last longer than four hours. The amount of snowfall is not included in the definition because the near zero visibility in blowing snow is the main concern. But even if severe winter storms do not meet the criteria of the official definition, they are not to be taken lightly. The winter storm referred to as the Great Lakes Blizzard of 1977 is still in the memory of many Ontarians.

Blizzards are a challenge to correctly forecast. The forecaster must know how much snow to expect and, more importantly, the snow's characteristics. For a blizzard, the winds must be able to pick up and blow loose snow. Sunshine and temperature changes can cause a crust to form on surfaces, which prevents the snow from blowing.

Snowsqualls

Snowsqualls or lake effect snow is the result of cold air blowing over a relatively warm body of water. As the air moves over the water, it picks up heat and moisture. Convective clouds develop along bands aligned with the mean wind direction through the cloud depth. Upon reaching land, the cloud bands encounter friction and are forced to ascend the upslope, onshore terrain. Snow falling from the convective clouds tends to be large dendritic flakes and accumulates rapidly, sometimes at the rate of 10 centimetres per hour. Much of this snow is dropped within 50 kilometres of the shoreline, with amounts diminishing as the snowsquall penetrates farther inland.

Storms

> *Snowsqualls are examples of severe convection associated with towering cumulus clouds that are only 5 kilometres tall, if that. By comparison, the cumulonimbus clouds associated with severe summer thunderstorms are sometimes 20 kilometres tall.*

In Ontario, snowsqualls are most likely from November to January in the areas east of Lake Superior, Georgian Bay and Lake Huron, which get the equivalent of 600 millimetres of precipitation in the form of light, fluffy snow from lake effect snowsqualls. The most intense snowsqualls are typically the earliest in the season, when the waters of the Great Lakes are still warm. Snowsqualls can also occur over smaller lakes, but the intensity of the resultant snowsqualls will be fetch limited. Sometimes southwesterly winds direct snowsqualls off Lake Ontario, which explains the peak in winter precipitation from Kingston to Brockville.

Snowsqualls are typically localized phenomena no more than a few kilometres across. Areas of subsidence and sunny skies separate the cloud bands. The spacing and height of the convective bands associated with snowsqualls are directly related to the height of the planetary boundary layer (PBL). The inland penetration of snowsqualls tends to be restricted overnight but can expand much farther inland during daytime hours, with the addition of daytime heating increasing the convective instability of the bands. Bands from Georgian Bay can penetrate to Ottawa and beyond.

Whiteout conditions are a major hazard with snowsqualls. The liquid-to-snow ratio can be as high as 1 to 20, and the relatively dry snow is easily blown by the winds associated with the snowsqualls. The convective winds are added to the general winds in the cold circulation and typically result in widespread blowing and drifting snow. The term whiteout is used to describe blizzard-like or blowing snow conditions that reduce visibility to just a few metres. People standing in a whiteout are unable to see shadows or landmarks and lose all sense of direction, perception and sometimes even their balance as the land and sky seem to blend into one. They cannot even see the hood of their own car.

Snowsquall prediction is intrinsically related to the wind direction, the peculiarities of the lake geography, the temperature of the water surface, the fetch of air over the water surface, the temperature and vertical stability of the cold air mass and the alignment of the winds in the PBL. Most of these quantities can be predicted with excellent accuracy days in advance. Although the precise location of the snowsqualls may still be in question, the occurrence of a snowsquall event is not. Forecasting an ongoing snowsquall is much more challenging because the storms tend to shift around with the ever changing winds. Even a subtle wind shift can redirect a snowsquall enough to change conditions from whiteout to sunny.

> *The heaviest 24-hour snowfall recorded in Ontario was 101.6 centimetres at Nolalu, near Thunder Bay, on March 24, 1975.*

Storms

Other Winter Storms

The following storm types are typical of El Niño years when the jet stream is predominantly from the northwest.

Alberta Clipper: a fast-moving low originating in the western Prairies that tracks southeastward across the Great Lakes at 30 knots or faster. This storm is a lower tropospheric manifestation of a rapidly moving short wave trough in the upper atmosphere westerly flow pattern. The disturbance typically moves eastward from the Gulf of Alaska, crossing the continent in just two or three days. When the upper level disturbance crosses the mountains, the associated surface low pressure system becomes somewhat disorganized. But in the lee of the Rocky Mountains, the upper level disturbance initiates the development of a low pressure cyclone in Alberta, which explains the name. An Alberta Clipper typically intensifies as it crosses the Great Lakes. As a result of its fast speed and relatively dry source region, it typically only delivers 2 to 5 centimetres of snow in three to six hours as it passes by.

Prairie Schooner: a relatively slow-moving low originating in the western Prairies that tracks southeastward across the Great Lakes at speeds of 15 to 25 knots. A Prairie Schooner is identical to an Alberta Clipper, but it moves slower and can bring double the amount of snow to Ontario. Typically, a Prairie Schooner delivers about 10 centimetres of snow in 12 hours as it passes by. Snow amounts can be even heavier if the Prairie Schooner is able to tap into a supply of moisture off the Gulf of Mexico. Heavy snowfall warnings may have to be considered in onshore flow areas where lake-enhanced snow can add to the system snow, resulting in totals that exceed the 15 centimetres of snow accumulation per 12-hour warning criteria.

Ice Boat: a very slow-moving cold frontal trough and low originating in the central Arctic that tracks southward across the Great Lakes at speeds of only 10 to 15 knots. It intensifies significantly as it crosses the Great Lakes. Snowsqualls off the Great Lakes are the main concern in the wake of Ice Boats because of their slow movement and low temperatures. The Arctic source region is very dry, so the heavy localized snowfalls that result are purely the result of snowsqualls and the very high liquid-to-snow ratios (1 to 20 or higher) in the arctic air mass. These patterns can also result in high wind chills as well as blizzard conditions along the Hudson Bay shoreline.

Summer Storms

Thunderstorms

Thunderstorms are an almost daily occurrence across Ontario in summer. They occur most frequently across southern Ontario (see Figure 8-4). Thunderstorms are the source of virtually all severe summer weather, including hail, torrential rain and flash floods, damaging winds and tornadoes. Even weak thunderstorms are a threat because of lightning. In fact, lightning kills more people than any other summer weather hazard.

> *The terms "Prairie Schooner" and "Ice Boat" were coined by the author, based on solid meteorological science.*

Storms

The storm that struck southwestern Ontario on July 14, 1997, is a classic example of severe summer convection. In mid-afternoon, the weather radar screens at Environment Canada showed no rain at all. Twenty minutes later, a severe thunderstorm formed near Punkeydoodles Corners, between Kitchener and Stratford. In just three or four hours the storm dropped more than 200 millimetres of rain. Flooding was severe. Winds of more than 115 kilometres per hour uprooted trees and flipped over small aircraft as far east as Guelph.

Ontario thunderstorms come in many modes of convection—pulse, multicell, supercell, high and low precipitation supercells, etc. To thoroughly discuss these would fill yet another book. As a generalization and especially in La Niña years, the severe late-spring and early-summer storms tend to be dominated by lines of supercells emerging from the southwest with a feed of Gulf of Mexico heat and moisture and a westerly to northwesterly jet stream. This pattern is common with La Niña and produces the big Ontario hail and tornado events. The supercell outbreak of April 1996 (1995–96 La Niña) was a classic and produced many severe events including the F3 Violet Hill tornado. The linear cold frontal supercell outbreak of Black Friday, May 31, 1985 (1983–85 La Niña), produced 11 tornadoes—including two F4s—as well as softball-sized hail, and it killed 12 people.

By mid-summer, strong daytime heating becomes the dominant forcing mechanism. In July, the sun is strong enough to elevate even pulse convection to severe limits. Single cell or pulse thunderstorms live through a life cycle lasting less than one hour. The sun heats

Storms

The back flanking line of a strong cumulonimbus

Average Annual Number of Days with Thunderstorms

Selected Cities	
Big Trout Lake	19
Kenora	24
Sioux Lookout	25
Thunder Bay	23
Kapuskasing	19
Timmins	19
Sault Ste. Marie	21
Sudbury	22
North Bay	24
Ottawa	24
Kingston	27
Toronto	28
London	31
Sarnia	28
Windsor	33

Fig. 8-4

the earth, which in turn heats bubbles of air that rise convectively. A cumulus cloud is born, which continues to grow into a towering cumulus (the cumulus stage) and, within the hour, a cumulonimbus thunderstorm (the mature stage). In the dissipation stage, the rainfall from the thunderstorm cools the ground, and the entire cloud is dominated by weak downdrafts and dissipating rainfall. It is often referred to as a "garden variety" thunderstorm because it typically causes no severe conditions except the threat of lightning, which is present with all thunderstorms.

In July, the pulse thunderstorm can explode with such upward buoyant energy, supplied by the sun and available summer moisture, that the associated weather can reach severe limits. This daytime heating is even more noticeable in the warmer El Niño years. Also, in El Niño years the severe supercell convection tends to shift farther north into northern Ontario, closer to the jet stream. The Pakwash Blowdown of 1991 (1990–93 El Niño) over northwestern Ontario (west of Lake Superior) was a high-precipitation supercell that destroyed 191,000 hectares of forest. The August 2, 2006 (2006–07 El Niño) supercell outbreak across central and eastern Ontario became one of the largest severe events in Ontario history, with 14 confirmed and 4 probable tornadoes and a total of 65 severe events detected. All events were predicted with long lead times, and thankfully—whether because of good luck or because the public practiced their safety and security plans—no one died.

Six to eight centimetre-diameter hailstone from Mildmay, Ontario July 8, 2007

Average Annual Number of Days with Hail

Selected Cities	
Big Trout Lake	1.3
Kenora	1.6
Sioux Lookout	1.4
Thunder Bay	1.0
Kapuskasing	0.7
Timmins	1.0
Sault Ste. Marie	0.4
Sudbury	0.9
North Bay	0.5
Ottawa	0.7
Kingston	0.4
Toronto	1.1
London	1.5
Sarnia	1.0
Windsor	1.3

Fig. 8-5

Hail and cars do not play well together.

Storms

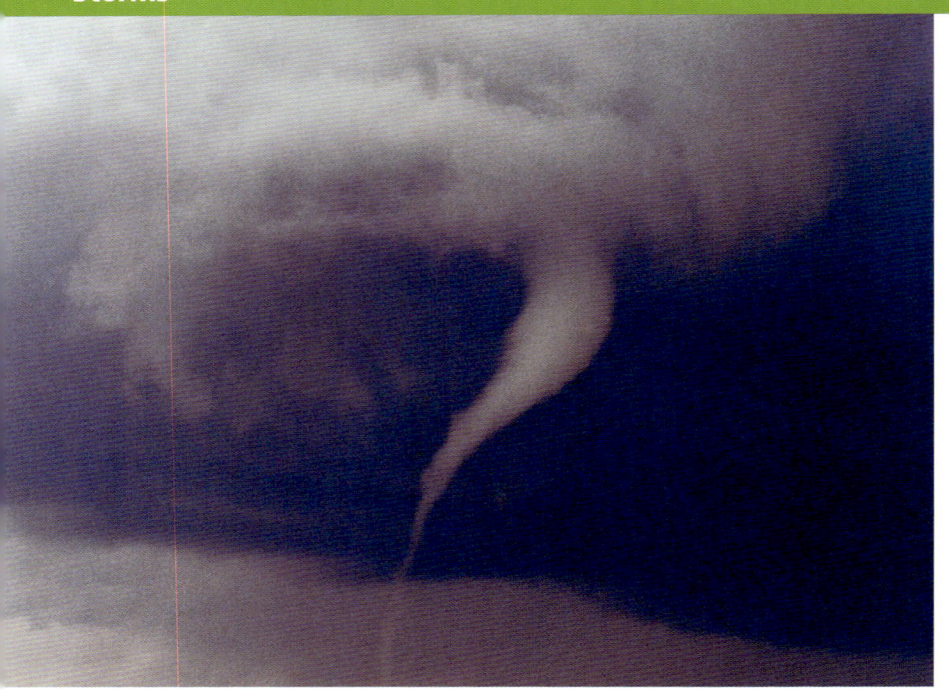

By early September, the convective season starts to wind down as the sun gets lower in the sky. Severe events do continue, though, and have been known to occur even into December.

Tornadoes

The fury of tornadic storms is unmatched by any other weather phenomenon. The rotational energy of the parent cumulonimbus cloud can be concentrated into just a few metres, and wind speeds can reach over 400 kilometres per hour. Based on the nature of the damage these storms inflict, few tornadoes in Ontario surpass F3 on the F-scale (see p. 93), though F4 tornadoes have been recorded.

Lake breezes play an important role in the creation of convection in Ontario. The main Ontario tornado alley stretches from London northeastward toward Barrie and Lake Simcoe. Lake breezes from Lake Erie and Lake Huron sometimes merge along this line, which is also influenced by the prevailing wind flow. In the probable southerly flow situation, this merger line will lie closer to the Lake Huron shore. The likely southwesterly jet stream associated with a severe convective outbreak will guide the developing supercells from their point of genesis along the lake breeze convergence line toward the northeast.

A point of land jutting out into the Great Lakes, like that at Point Clarke, will result in similar but smaller lake breeze convergence lines that collide. Supercells created by this collision typically track northeastward toward Owen Sound along a secondary tornado alley. Similar shoreline shape effects can be found along the Lake Huron and Lake Superior coasts as well.

Storms

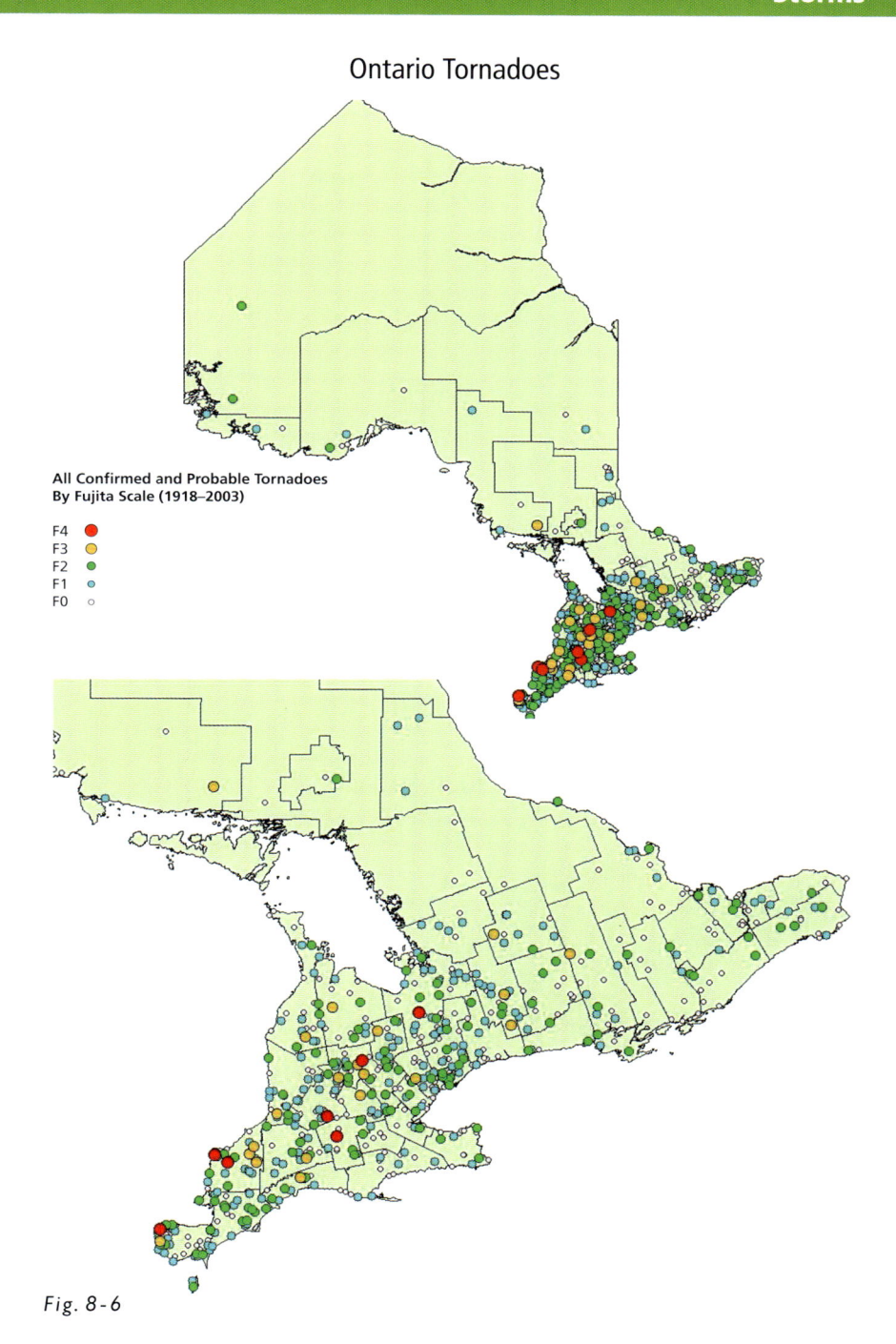

Ontario Tornadoes

**All Confirmed and Probable Tornadoes
By Fujita Scale (1918–2003)**

- F4 ●
- F3 ●
- F2 ●
- F1 ●
- F0 ○

Fig. 8-6

Storms

Average Annual Frequency of Tornadoes

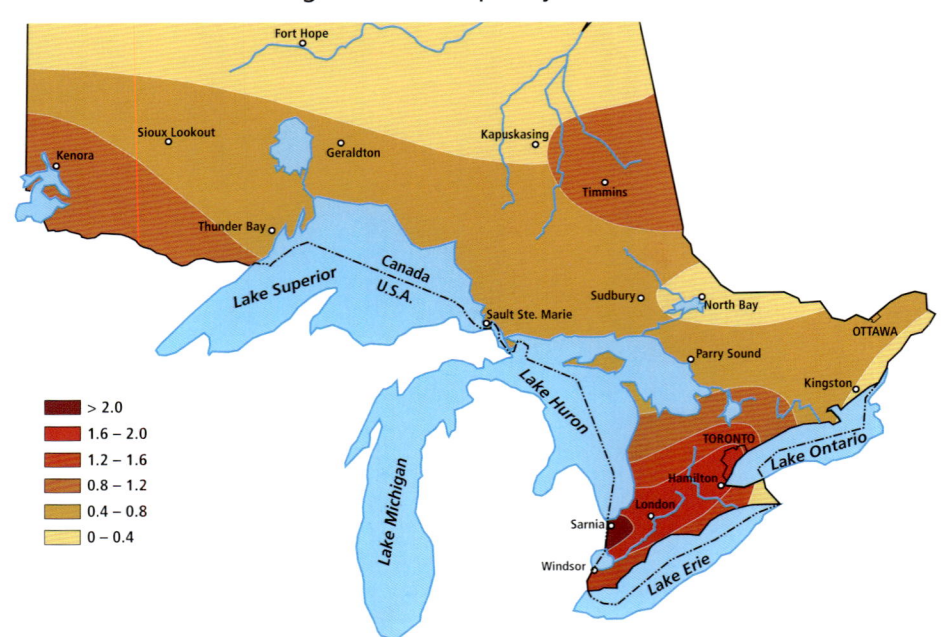

Legend:
- > 2.0
- 1.6 – 2.0
- 1.2 – 1.6
- 0.8 – 1.2
- 0.4 – 0.8
- 0 – 0.4

Fig. 8-7

August 2, 2006 Tornadoes

- F1 Marsden Lake
- F1 West Guildford
- F1 Mink Lake
- F2 Combermore
- F0 Clayton
- F2 Bancroft
- F1 Uffington
- F1 Fergusons Corner
- F1 Minden Hills
- F0 Catchacoma
- F1 Drag Lake
- F0 Kirk Cove – Coxvale
- F0 Myers Cove
- F1 Anstruther Lake

Fig. 8-8

158

Storms

F2 house damage, June 25, 2009

A typical Ontario tornado moves at 30 to 50 kilometres per hour toward the east or northeast. Its motion is fairly slow and predictable, allowing time for evasive actions. A cool northwestly wind indicates that the tornado has passed. The length of time a tornado is on the ground is highly variable. The parent storm complex often generates a series of tornadoes.

The number of tornadoes varies widely from year to year, and the long-term annual Ontario average is about 15. More tornadoes certainly occur, but go undetected. Ontario is one of the most tornado-prone provinces in Canada.

On May 31, 1985, at least seven separate tornadoes developed along a cold front as it swept across southern Ontario. The tornadoes were named after the communities they affected—Arthur, Grand Valley, Tottenham and Barrie. The Grand Valley and Barrie tornadoes reached F4 intensity. There were 12 fatalities, and property losses exceeded $100 million. On August 2, 2006, the most severe convective day in Ontario history spawned 18 confirmed and probable tornadoes and 65 severe events. Thankfully, no one died. Severe convection is a part of Ontario, and, unfortunately, history will repeat itself. Know what to do to protect yourself and your family.

Storms

Autumn Storms

Autumn is a beautiful season in Ontario, but when it turns nasty, it really turns nasty. The extratropical remains of hurricanes can be devastating to southern Ontario. Normally these are just huge rain producers, bringing 100 or more millimetres of rain along their path, but they can also be accompanied by damaging winds. Hurricane Opal of 1995 (1995–96 weak La Niña) brought 80 millimetres of rain along the lower Great Lakes. Rainfall accumulation in centimetres per 24 hours can be quickly estimated by dividing 240 by the speed of the hurricane in knots.

On October 15, 1954, the extratropical remains of Hurricane Hazel, the most famous hurricane in Canadian history, struck southern Ontario with winds up to 110 kilometres per hour and more than 200 millimetres of rain in less than 24 hours. In Toronto, bridges and streets were washed out, and homes and trailers were washed into Lake Ontario. Thousands of people were left homeless, and 81 people were killed (35 on one street alone).

Early-season Colorado lows can also be significant as they intensify over the Great Lakes. An example of this was the November 10, 1975 (1973–76 La Niña), storm that sank the Edmund Fitzgerald in Lake Superior, killing the entire crew.

Most of the precipitation that falls in Canada draws its water from the Pacific Ocean, the Gulf of Mexico and the Caribbean Sea.

Three Types of Weather Alerts

Special Weather Statements
- issued for unusual weather events that could inconvenience or alarm the general public and that are not described in the weather forecast
- may be issued for events that are occurring outside of but may make their way into the forecast region
- updated as needed

Watches
- issued when conditions are right for a potentially disruptive storm, but when the track and strength of the storm are unknown
- in summer, may be issued up to six hours before the event
- in winter, issued at least 12 to 24 hours before the event

Warnings
- issued when severe weather is occurring or will occur
- most are issued 6 to 24 hours in advance, except thunderstorm warnings, which may be issued less than one hour before the event
- are updated at least every six to eight hours or as needed

Fig. 8-9

Chapter 9: Observing the Weather

> **Old weather saying:**
> *Rain before seven,
> fine by eleven.*

What will the weather be like tomorrow? Next Friday? This time next year? These are questions that we ask. Looking out the window gives us our first impression as to what the weather will do for the next few hours, but if we are interested in what is going to happen tonight or tomorrow, looking out the window is inadequate. In the 19th century, when people first undertook an intensive, organized study of the atmosphere, they realized there could be very different conditions at two locations, even when the locations were not far apart. This realization led to the understanding that weather is organized into patterns and that these weather patterns move. So it was concluded that observations should be taken at many locations at about the same time and then gathered together at a central location where the patterns could be analyzed. From this analysis a forecast of future weather conditions could be prepared based on an understanding of the rules or laws that determine how the atmosphere works.

Observing the Weather

The nature of weather and how it behaves on different scales is based on the principles of fluid dynamics. The atmosphere is a fluid that is in turbulent flow on a rotating sphere. Mathematician Lewis Fry Richardson may have summarized it best when, early in the 20th century, he described turbulent fluid flow: "Big whirls have little whirls that feed on their velocity, and little whirls have lesser whirls and so on to viscosity." This ditty recognizes some basic realities of the atmosphere: that it is a turbulent fluid that behaves on many different scales of motion, and that there is a link between these scales of motion.

So, the essence of weather forecasting is to measure the smallest scales reasonable and use the rules of atmospheric motion that define the link between these scales to give a picture of the state of the fluid at some time in the future. Weather patterns have a space scale of a few hundred kilometres, and they can change over a few hours. To resolve these weather patterns requires a network of stations spaced about 100 kilometres apart that take observations every hour. This space and time scale is common among most meteorological observing programs that support weather forecasting.

Because measurements of weather parameters are taken from all over the world, the information is recorded in a special code that can be understood by people of all countries, regardless of the language they speak.

Observation Systems

Assembling observations and preparing forecasts for even a few locations requires a large-scale, coordinated approach. Consequently, the observing and forecasting systems are largely the responsibility of national organizations. The data-observing approach must be standardized to make the international exchange of data run smoothly. Observational standards follow an international protocol established and maintained by the World Meteorological Organization, which operates under the United Nations. The data is communicated to one or more locations, where it can be analyzed so forecasts can be prepared.

Weather forecasting requires data from the full depth of the atmosphere. In Canada, the surface weather observation network consists of several hundred stations across the country that repeat measurements, generally on an hourly basis. In addition, about 30 Canadian stations measure meteorological variables above the surface to heights of about 30 kilometres. Around the globe, 90 countries operate a total of about 1000 such upper-air observation stations. At these aerological stations, surface weather data is also measured, and many other specialized instruments are used for specialized observations that support atmospheric research programs. These surface and upper-air *in-situ* measurements are supplemented with data from remote sensing systems such as weather radar and weather satellites. The data is sent to local and national weather offices, and much of it is made available to the public through the internet and the media.

Observing the Weather

Gathering marine weather data

Surface Weather Observation

Formal weather observations were recorded at locations across British North America through the 18th and early 19th centuries. The first official observation program began on Christmas Day in 1839. According to a plaque on the University of Toronto campus, the British Army began regular meteorological and magnetic observations on the site, which was then called Her Majesty's Magnetical and Meteorological Observatory. John Lefroy, the observatory's director, travelled west in 1843 to make magnetic observations in the Hudson's Bay Company's territories. The first weather observation network was established in Great Britain in the 1850s. Canada's first network was established in 1872. It covered southern Ontario, Québec and the Maritime provinces, and data was communicated using the newly installed telegraph systems. During the 1880s, the network expanded into what is now Alberta, following the Canadian Pacific Railway and its telegraph network. At present, the Meteorological Service of Canada operates the national weather observation program that supports the preparation of weather forecasts.

The measurement of atmospheric variables has a long history, and electronic technology and computerization have influenced weather observing methodologies. Meteorology has evolved into several areas of specialization, and the monitoring requirements have evolved as well. Data requirements for weather forecasting are somewhat different than those needed for atmospheric research or for specialized applications in air quality. Weather forecast requirements are based on what are known as standardized observing protocols. Issues including instrument location, representativeness and accuracy of measurement, and observation averaging times must

Observing the Weather

be considered if data is to be used in the preparation of forecasts. Predicting the atmosphere is difficult enough without the need to take into account idiosyncrasies for each weather measurement.

At weather observing stations, the instruments are located where they provide a measurement that is representative of the atmosphere. For example, they are not sheltered by a tree or building, a temperature reading is not artificially elevated by direct sunlight or the heat rising off a paved surface, snow measurement is not subject to a snowdrift depending on wind direction, and a precipitation measurement is not elevated by rain dripping off a roof or swirling around a tree.

Most observations are from unstaffed sites where data is measured automatically. Many sites are located at airports, where trained observers provide additional data that is useful for the aviation industry.

Many instruments are housed in a ventilated cabinet known as a Stevenson screen. Named after its inventor Thomas Stevenson (father of the famous author Robert Louis Stevenson), the cabinet ensures that sunlight does not fall directly on temperature gauges. The cabinet is painted white to ensure that the sun's rays do not cause the cabinet to heat disproportionately.

Each Stevenson screen contains different types of thermometers.

Observing the Weather

Precipitation

The history of precipitation measurement goes back to the 4th century BC in India. The principle used then is the same one used now—measure the amount of precipitation collected in an open-topped container. Refinements have been focused on reducing the loss of precipitation from splash-out of raindrops, reducing the effects of wind on collection efficiency of the opening, and reducing evaporative loss between precipitation events. The manual rain gauge used in Canada has a design that has been largely unchanged since 1871. The rate of rainfall is measured by a tipping bucket rain gauge. Collected rain passes into a small container that empties each time its capacity is reached. The rate at which the container tips, or empties, indicates the rainfall rate.

Snow measurement has proven to be a challenge. The snow catch rate of a collection container is greatly influenced by wind, so most devices have a shield to reduce wind speed around the opening. The shape of the opening also influences how efficiently a container collects snow. The Canadian Nipher-shielded Gauge has a bell-shaped opening to minimize loss. Where there is a human observer, the depth of snow accumulated on the ground is determined by a simple depth measurement at a number of sites in a so-called snow course. Measurements are taken at several locations so that differences in accumulation that result from wind drifting can be averaged.

A tipping bucket rain gauge (above); the tipping bucket (below)

Observing the Weather

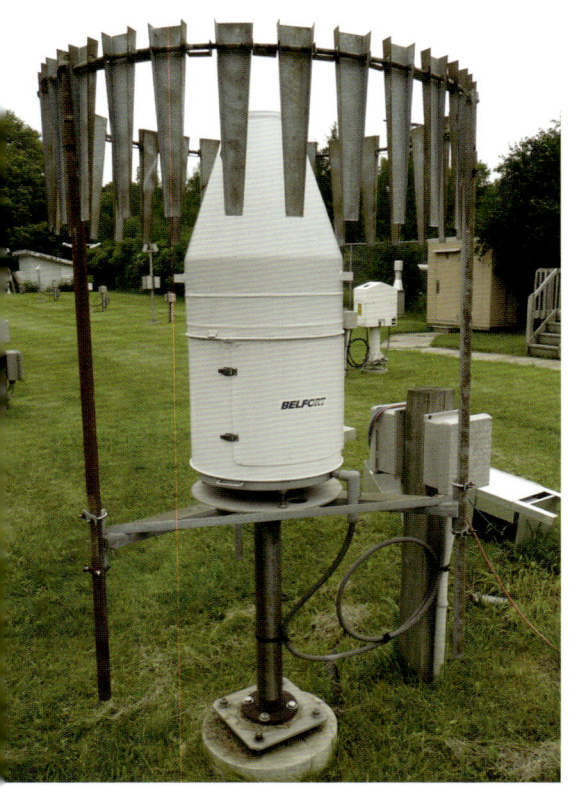

Nipher Snow Gauge

Many innovative techniques have been used to measure snowfall automatically. Snow pillows laying on the ground measure the weight of the snowpack that accumulates during winter. The storage gauge collects snow in a large container filled with glycol, which melts the collected snow, and measurements of the weight of the container are taken periodically. The depth of snow can also be measured at one point on the ground by a downward-pointing microwave sensor.

Automated detection that identifies the type of precipitation is difficult. One approach measures the fall rate of the precipitation by means of a small Doppler radar and links the fall rate to the type of hydrometeor. This approach works well enough to differentiate between snow and rain, but discriminating between drizzle and snow is not easy because they have essentially the same terminal velocity. The detection of freezing precipitation is also a challenge. One approach uses a vibrating probe and calibrates the rate of vibration to the rate at which precipitation adheres.

Temperature

Around 1592, Galileo invented the first thermometer, which used the expansion of air to measure temperature. The problem was, this expansion-of-air approach was prone to many errors, and the apparatus was large and cumbersome. Scientists soon realized that thermometers made of sealed tubes containing mercury or alcohol were much more reliable and practical.

Gabriel Daniel Fahrenheit developed a temperature scale in 1724 that had three fixed reference points. Zero was the value given to the coldest temperature that he could create by mixing ice, water and ammonium chloride. The upper end of the scale was the temperature of the human body, which Fahrenheit arbitrarily called 96 degrees. The third fixed point was the temperature of a water/ice mixture that, on his scale, had a value of 32 degrees. The Celsius temperature scale, invented by the Swedish astronomer Anders Celsius in 1742, is the one most frequently used worldwide. On this scale, the temperature of the water/ice

Observing the Weather

Maximum and minimum thermometers in a Stevenson screen

mixture was set as zero, and the value of 100 degrees was assigned to the temperature at which pure water boils.

Various types of thermometers are used at surface weather stations. The minimum thermometer contains a tiny metal dumbbell (index) that floats in liquid, usually alcohol. The index is forced toward the bulb by the retreating surface of the liquid as the temperature falls. When the temperature rises, the dumbbell remains in place and registers the minimum temperature. The maximum thermometer has an index that is pushed upward in the barrel of the thermometer and registers the maximum value reached in the observation period. Regular, maximum and minimum thermometers are all manually read.

Temperature sensors have been developed for use in automated systems or where space limitations or working environments are unsuitable for liquid-filled thermometers. The most common are instruments known as thermistors (the name is a combination of "thermal" and "resistor"), which are made of mixed metal oxide semiconductors. In a semiconductor, the resistance to the flow of electrical current changes as the temperature of the device changes. In some thermistors, the resistance increases with increasing temperature, and in others, the resistance decreases with increasing temperature. The semiconductor devices can be physically small and rugged in design, which makes them suitable for use with electronic circuitry.

Observing the Weather

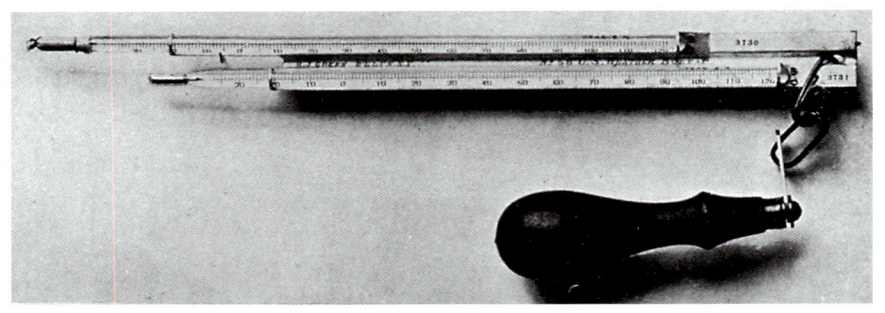

A sling psychrometer to measure humidity

Humidity

Humidity measurements are important in meteorology because knowledge of the moisture content of air is necessary to determine the formation of cloud. In 1660, Francesco Eschinardi demonstrated that the evaporation of water causes a thermometer to cool. The psychrometer is a device that uses this principle to determine the amount of moisture in the air based on the difference in temperature between two thermometers. The dry bulb thermometer measures air temperature normally, whereas the so-called wet bulb thermometer has its bulb enclosed by a water-saturated wick. A fan draws air past both thermometers at a known rate, and the temperature difference is then used to determine the humidity based on previously calibrated values.

The psychrometer is not suitable for automated measurements of atmospheric humidity because the water reservoir would need to be maintained, and the behaviour of the device changes at temperatures below freezing. An instrument called the dewcel is used at Canadian weather stations. This device has a fibreglass sleeve that is saturated with a lithium chloride solution. The lithium chloride absorbs moisture from the air, and a heater is used to raise the temperature of the sleeve, causing the moisture to evaporate. A thermistor measures the temperature difference between the heated and unheated sleeve. This temperature difference is proportional to the atmospheric moisture.

Many other devices have been developed to measure humidity. They use different principles to measure moisture, such as the change in length of hair as it absorbs water vapour, the change in electrical properties of materials as they absorb water vapour, or the temperature at which a cooled reflective surface becomes fogged because of water vapour.

Canada's Alert weather station is the world's most northerly. It is located on Ellesmere Island, 830 kilometres from the North Pole.

Observing the Weather

Pressure

Atmospheric pressure is a result of the weight of overlying air. Surface pressure has traditionally been measured by the mercury barometer (meaning "weight meter"), which Evangelista Torricelli invented in 1644. The modern instrument consists of a mercury-filled tube that is inverted and immersed in a mercury-filled cistern. Atmospheric pressure pushes down on the top of the cistern and supports the column of mercury in the tube. The height of mercury that is supported essentially balances the weight of the atmosphere above the cistern. At sea level, this amounts to the weight of about 76 centimetres of mercury. This type of atmospheric measurement is the most consistent and reliable available. However, the mercury barometer is not easily transportable, it does not lend itself to automation, and there are health concerns associated with using mercury. As a result, the aneroid barometer, invented by Lucien Vidie in 1843, is widely used.

Mercury barometer at an airport (above); aneroid barometer with graph to record pressure changes (below)

Observing the Weather

The aneroid barometer consists of a partially evacuated and sealed metal bellows canister that expands or contracts in response to atmospheric pressure changes. Mechanical gears or electronic techniques determine the amount of this deflection.

Aneroid systems are used in portable barometers, altimeters and pressure recording devices (barographs). Other pressure measurement techniques using quartz and silicon-based cells are used in specialized applications. To measure atmospheric pressure, the barometer must be located in a building where the temperature is controlled for the stability of the instrument. To ensure the pressure represents outside conditions, the building must not be sealed, and it must be designed to minimize the effects of spurious pressure fluctuations caused by wind pressure, vehicles moving nearby and even the opening and closing of doors.

Wind

Wind may be the most influential weather phenomenon throughout the ages. The Tower of the Winds in Athens was built by Andronicus of Cyrrhus around 50 BC to recognize the eight quadrants from which the wind blows. A weather vane was mounted on top of the tower. Since that time, various types of wind vanes have been developed to measure wind direction, and several different methods have been used to measure the magnitude of the wind. One early approach attempted to measure wind force directly using a suspended flat plate that was deflected by the wind. Such devices date back to about 1450. Subsequent variations on this theme consisted of a flat plate, either square or circular, which was kept facing the direction of the wind by a vane. One such device invented by A.F. Osler was installed at the observatory in Greenwich, England, in the mid-19th century.

A more useful measurement concept is wind speed—the distance that air moves over a short period of time. Wind speed is often measured with the cup anemometer, which operates on the principle that cups mounted on arms rotate around a vertical axis as the air moves past. The number of rotations in a fixed interval of time is proportional to the wind speed. John Thomas Romney Robinson invented the first such instrument in 1846, and Canadian John Patterson invented the common three-cup anemometer in 1926. The anemometer and wind vane are generally mounted on a tower that meets an international standard for height and siting. Measurements are taken 10 metres

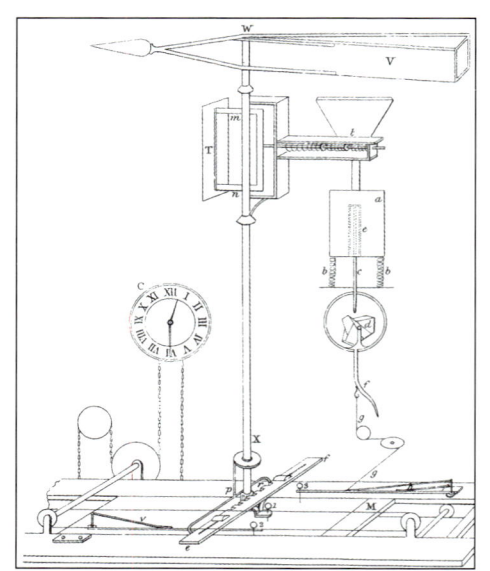

Osler's wind measurement device

Observing the Weather

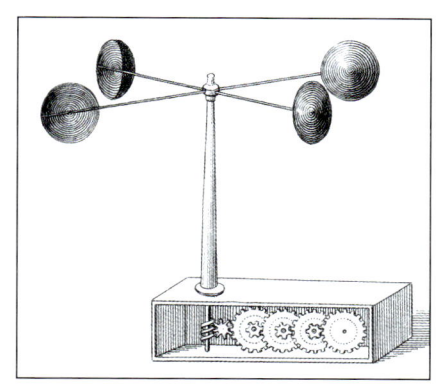

The Robinson cup anemometer

above ground, and the tower is located where the wind flow will not be influenced by any trees or buildings.

Other wind measurement approaches have been developed for special applications. At many automated weather stations, the wind speed and direction devices are combined into one instrument. This device has a propeller mounted at the front of a wind vane. Non-mechanical devices can also be used to measure rapid fluctuations in wind speed and direction. Instruments known as hot-wire and hot-film anemometers measure the cooling effect of air as it moves across an electrically heated surface. A stronger wind removes more heat and, therefore, more electrical energy is required to heat the surface. Yet another approach measures the pressure drop that occurs when wind flows across a small opening. The pressure tube approach was first developed in the late 18th century, and a practical version developed by William Dines in 1892 is still used today. This approach measures the speed at which an aircraft moves through the air, but it is also used in ground-based instruments to measure rapid fluctuations in wind speed. The sonic anemometer calculates wind speed by measuring the time taken by two sound pulses to move in opposite directions across a known distance. The difference in time that the two pulses travel is proportional to the speed of the air. This instrument has the advantage of no moving parts as well as being able to measure rapid changes in wind speed.

Jet inbound to Pearson International Airport past the national headquarters of the Meteorological Service of Canada (showing standard wind vane with Patterson three-cup anemometer).

171

Observing the Weather

A propeller-and-vane anemometer

Visibility

This parameter is important for such things as aviation and marine transportation. Visibility is a function of atmospheric conditions, and it also depends on the capability of the human eye, making it one of the more subjective atmospheric measures. A number of instrumentation methodologies have been developed to bring a degree of objectivity to this parameter. Most instruments measure the drop in intensity of a light source when observed over a fixed distance in the atmosphere. These instruments are used at airports to measure the runway visual range, which is critical for aircraft movement around airports.

Observations at Sea

A major international cooperative program is in place to gather weather information from over the oceans. Observations taken on vessels use much of the same types of instrumentation that are used at land stations. However, because the ships are moving, the wind measurements must take into account the vessel's motion. In addition, the ideal siting criteria used at land stations cannot be met in the cramped conditions onboard most vessels. Still, the key measurements of temperature, pressure and wind are still useful in the weather forecasting programs.

Marine buoys are the oceanic version of the unstaffed automatic land stations. There are two types of buoys—moored and drifting. The moored buoys have a long tether fixed to the ocean floor. The observation program usually includes pressure, temperature and wind as well as marine measurements (such as sea surface temperature) and sea state information (such as wave period and height). Occasionally, the buoys also collect ocean current information. Drifting buoys are generally limited to temperature and pressure measurements, but some gauge wind speed as well. Data is usually communicated through the meteorological satellites to the world communication network. Although the location of the moored buoys is known, the drifting buoys must be tracked using global positioning technology. The motion of the drifting buoys provides a measure of the ocean current.

Observing the Weather

Measurement Aloft—the Aerological Monitoring Program

Weather forecasting programs need regular measurements of the upper atmosphere. These aerological measurements, as they are known, are taken year-round every 12 hours, at noon and midnight Coordinated Universal Time (that's 7:00 and 19:00 Eastern Standard Time in Ontario). The spacing of the upper-air observing network is much wider than the surface observing network, and stations are about 600 kilometres apart. In Ontario, two stations in the national network are located at Moosonee and at Pickle Lake. Measurements are taken using a lightweight instrument package that is lifted using a helium-filled balloon. The instrument package, known as a radiosonde, includes sensors that measure pressure, temperature and moisture. It has a small radio transmitter onboard that sends the data and also allows the motion of the instrument to be tracked at the receiving ground station. The horizontal distance moved by the instrument package during its ascent is used to calculate the wind speed and direction. The balloon bursts at a height of about 30 kilometres above ground level, and a small parachute slows the instrument package in its fall back to earth. These instrument packages are not normally recovered. The balloon-borne observations are taken in all weather conditions. It is particularly challenging to release the balloon when the surface winds are strong and freezing rain or blowing snow is occurring.

Once a week, ozone measurements are taken at 10 Canadian aerological sites. This monitoring program focuses on ozone in the stratosphere and the depletion of this layer, which protects life forms from the harmful effects of ultraviolet radiation. At present, the Egbert site just north of Toronto releases this ozonesonde. Ground-based instruments also measure ozone with remote sensing techniques.

Although the aerological network is confined primarily to land stations, there are a few ships that release radiosondes at fixed locations on the oceans. These programs are expensive to operate, and they are now being replaced by programs operated from ocean-going transport vessels, which release radiosondes while en route.

A visibility sensor

Observing the Weather

Weather Radar Interpretation Errors
(adapted from Environment Canada website)

1. Blocking Beam

Uneven terrain such as hills and mountains can block a radar beam, leaving gaps in the pattern—this is very common in the mountains.

2. Beam Attenuation

Storms closest to the radar site reflect or absorb most of the radar energy, so little energy is left to detect storms that are farther away.

3. Overshooting Beam

When clouds are close to the ground, as they are in lake effect snowsqualls, the radar beam may overshoot them so that even clouds with intense precipitation will only have a weak echo.

4. Virga

The radar beam detects precipitation that is occurring aloft, but the precipitation never reaches the ground because it is absorbed by low-level dry air conditions.

5. Anomalous Propagation

During an inversion in the low atmosphere, when a layer of warm air overlies a layer of cooler air, the radar beam is not able pass between the two layers and is reflected back to the ground, sending a false strong signal back to the radar site—this event is most common in the morning as well as in spring over the cold waters of the Great Lakes.

6. Ground Clutter

The radar beam echoes off objects on the ground, such as tall buildings, trees or hills.

Remote Sensing

Weather observations using instruments that are in direct contact with the atmosphere are known as *in-situ* measurements because they have sensors in direct contact with the atmosphere. In contrast, remote sensing techniques probe the atmosphere from a distance.

Various techniques are used for remote sensing, including sound, the reflection of electromagnetic waves (from sunlight and microwave and laser instruments) and the sensing of heat radiated by a substance.

Weather Radar

When radar (an acronym for radio detection and ranging) was first used to track the motion of aircraft, the returns from clouds and precipitation were considered detractive noise. However, it quickly became apparent that radar could be used to provide new information about the atmosphere. Weather radars have a transmitting dish that rotates and sends pulses of radio waves at microwave frequencies. The dish focuses the pulses along a narrow beam that reflects off cloud and water droplets. When part of the microwave pulse bounces back to the antenna, the radar's electronics time the return and measure its intensity. One-half of the total time for the pulse to travel to and return from the target multiplied by the speed of light determines the target's distance from the radar dish.

Observing the Weather

The strength of the reflected signal is proportional to the size of the reflecting object. Microwave radiation reflects off cloud and rain droplets (and even swarms of insects and birds). Based on the physical properties of raindrops, the intensity of the returned signal is proportional to the size distribution of the reflecting droplets. Using rain gauge data, the rainfall rates can be correlated with the assumed droplet size. In this way, the rainfall rates in the sensed clouds can be estimated. Canadian weather radars transmit using a 5-centimetre wavelength. They are configured to have an effective viewing distance out to about 250 kilometres. The radar display is a circular pattern centred at the radar site. Usually a sequence of images is displayed, which then shows the motion of the echo. The Canadian weather radar network is operated by Environment Canada and includes eight radars in Ontario, but parts of Ontario are also covered by radar in Manitoba, Quebec and the U.S.

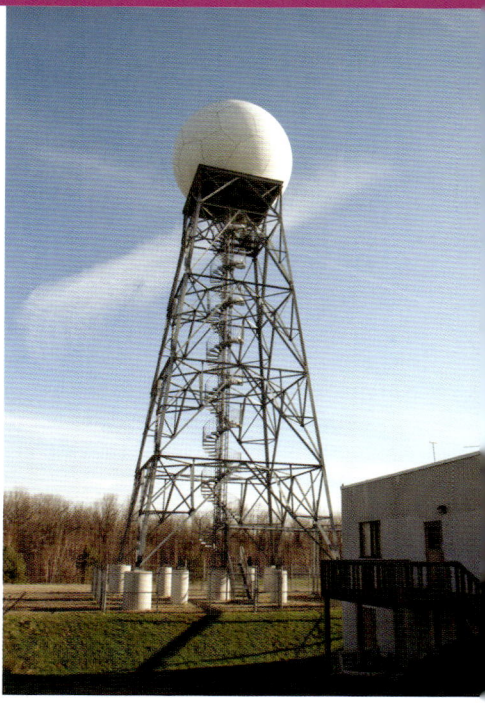

King City radar site (above)

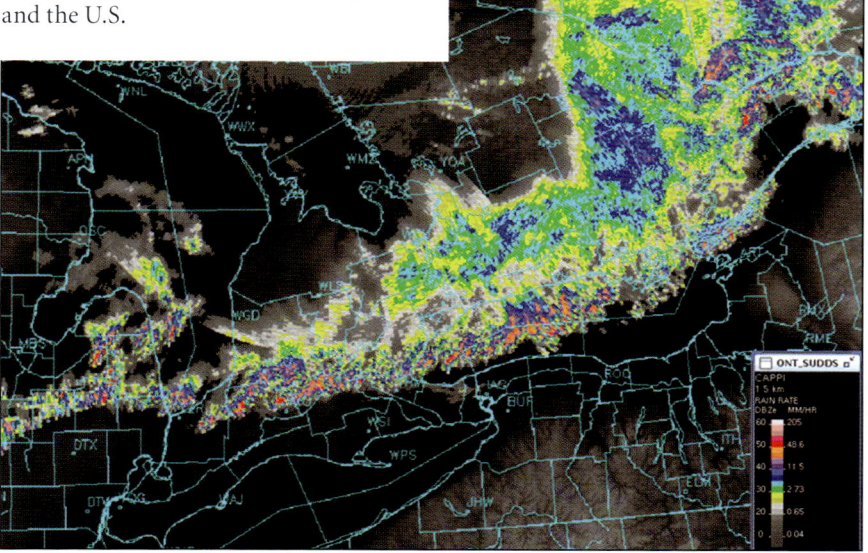

The most severe convective day in Ontario history—August 2, 2006

Observing the Weather

All Canadian radars (31 in 2008) also make use of the Doppler principle to measure the shift in frequency of the returned echo. Changes in the frequency of the returned signal are proportional to the speed of the precipitation either directly toward or directly away from the radar site. There is no frequency shift produced by precipitation moving perpendicular to the radar beam. This Doppler principle is used to measure the flow patterns in the atmosphere. It is useful for examining the wind patterns in intense thunderstorms, and the Doppler radar can even be used to determine if the cloud echo is rotating. A rotating cloud echo indicates a potentially severe thunderstorm, which may generate a tornado even though tornadoes themselves are below the resolution of the radar beam. The Doppler signal is also useful in diagnosing freezing rain events and the motions of wind shift lines and fronts.

Another radar system, which uses alternating pulses of horizontally and vertically polarized signals, has been developed. This dual polarized radar system, also referred to as polarimetric weather radar, has several advantages over conventional radar in estimating precipitation types and amounts. Polarimetric weather radar can discriminate between hail and rain, detect mixed phase precipitation and estimate rainfall volume. The information is quite complex, and computerized precipitation classification schemes are employed to process the radar information to estimate the characteristics of the hydrometeor. The King City weather radar station north of Toronto has been operating a polarimetric weather radar in research mode since 2006.

The Doppler Effect

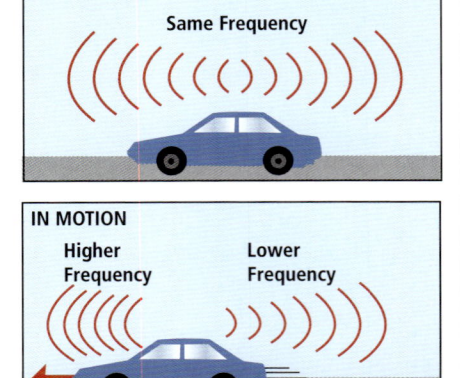

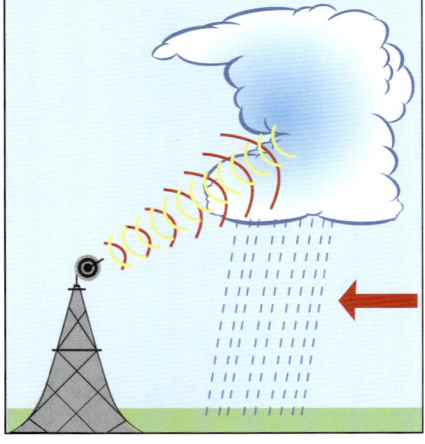

Fig. 9-1 The frequency of sound waves is shifted because of the motion of the source. The frequency of microwave radar echoes is shifted because of motion of cloud and rain.

Observing the Weather

Weather Satellites

The weather satellite is a relatively recent invention but has become one of the most useful tools for observing large portions of the globe. The TIROS satellite launched by the United States in 1960 was the first, followed by scores of satellites that have since been launched by governments and other agencies.

Most weather satellites use passive sensing systems, which receive two basic types of radiant energy. Visible light produced by the sun is reflected off the earth's surfaces and clouds, back up to the satellite. Visible sensors on meteorological satellites are essentially black and white cameras. Clouds look white to the satellite-based sensor, and ground surfaces and water bodies appear grey or black. A second type of sensor detects infrared or heat energy. The intensity of the infrared energy is related to the temperature of the emitting surface, a phenomenon known as Boltzmann's law. The satellite infrared sensors are sensitive to the temperatures of earth's land and water surfaces as well as cloud tops—a temperature range extending from -70° to 60° C. Because the earth and atmosphere emit heat day and night, infrared images are always available, unlike the visible images, which are only available during daylight.

There are two types of satellite systems now generally in use. Geostationary satellites are placed in circular orbits over the equator at heights of 35,800 kilometres. The satellite moves in the same

Weather Satellites

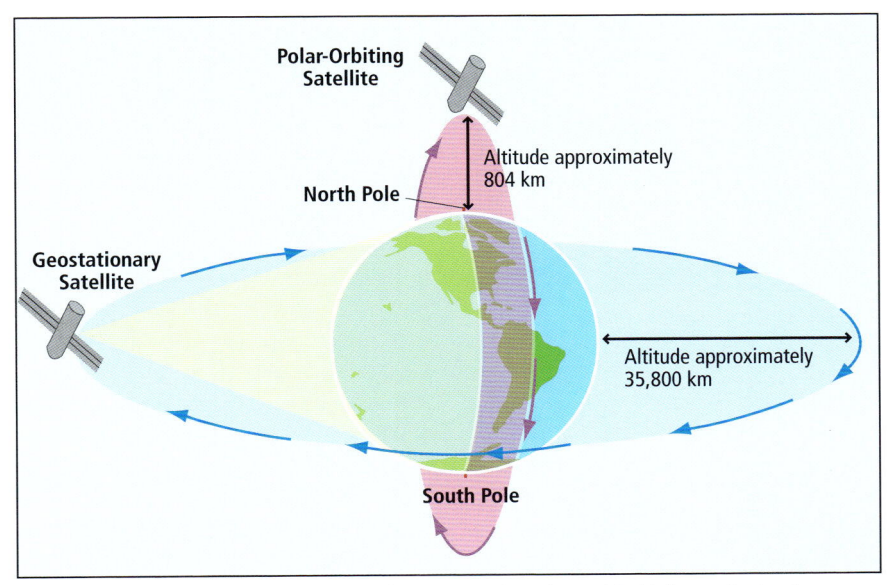

Fig. 9-2 Polar-orbiting satellites are close to earth and provide high resolution images. Geostationary satellites are in high orbits and provide low resolution images.

Observing the Weather

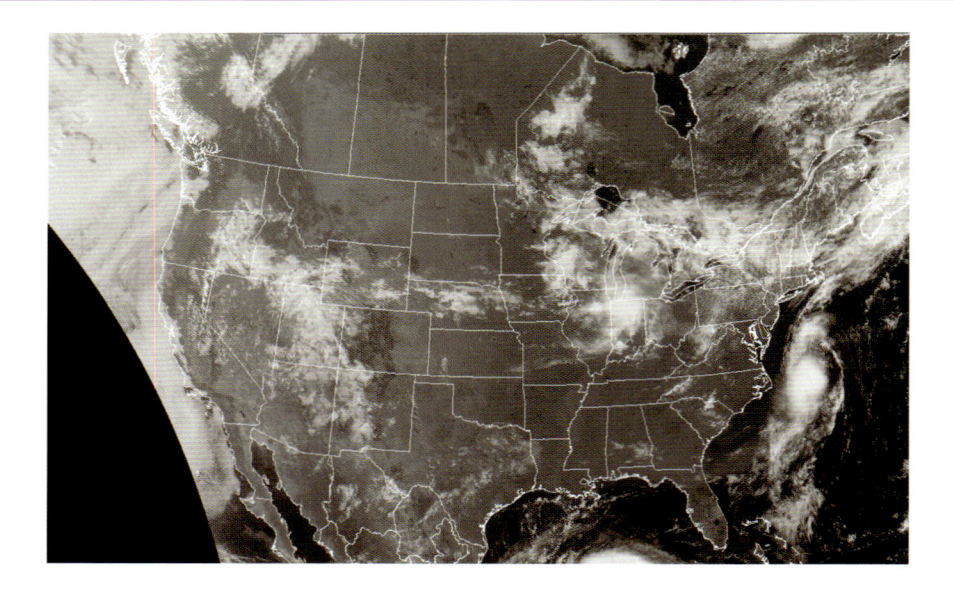

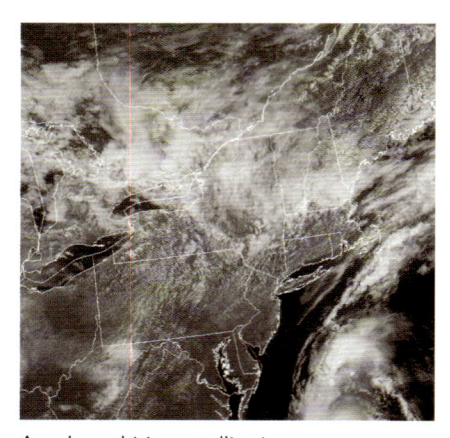

A polar-orbiting satellite image

direction as the earth rotates and takes one full day to orbit the earth. It is essentially stationary above the earth's surface. The second type of satellite is in a much lower orbit at about 800 kilometres above the surface. The orbit passes near the poles and is at a slight angle of inclination relative to the equator. The orbital geometry is such that the satellite orbital plane moves in time with the sun. These polar-orbiting satellites take a continuous swath of images along their orbit and, as a consequence, the images appear as one long, continuous image. Because geostationary satellites are about 45 times higher than polar-orbiting satellites, the resolution is lower, and images from the equator-oriented satellites become greatly distorted toward the north and south poles. The polar-orbiting satellites move around the earth in about 100 minutes, and most places on the earth's surface are scanned twice daily, once in daylight and once in darkness. Because they pass frequently over the poles, these satellites provide more useful data at those latitudes than do the geostationary satellites. The infrared sensor on geostationary weather satellites can distinguish areas that are 4 kilometres in width, whereas the images in the visible light spectrum can resolve objects as small as 1 kilometre.

Observing the Weather

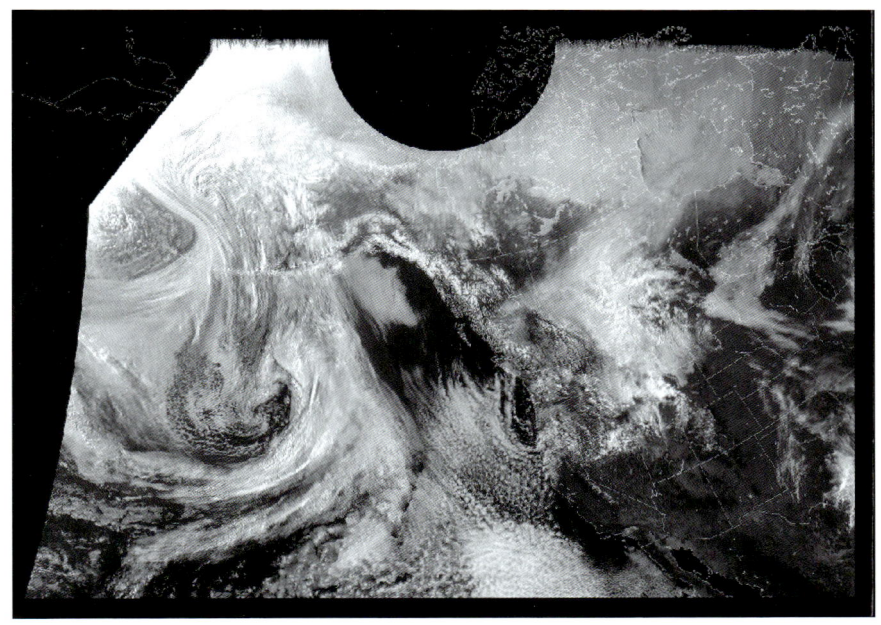

A geostationary satellite image

Satellite observations have an obvious advantage over areas of the earth for which there is little conventional weather data available. Tracking weather systems as they move across these areas has proven to be immeasurably valuable. Satellites are, in fact, used to produce "synthetic soundings" of the atmosphere over ocean areas where there are few aerological stations.

Acoustic Sounders

These instruments, called SODARs (for sonic detection and ranging), use a pulse of sound to measure wind in the lower atmosphere. Their principle of operation is similar to that of radar except that they use sound waves rather than microwaves. When a sound pulse meets a discontinuity in the atmosphere, sound is scattered in all directions. The discontinuity can be in the temperature field, such as a strong temperature inversion. There can also be discontinuities in wind, associated with eddies that are present in the turbulent air flowing over the earth's surface. The SODARs' speaker directs a burst of sound, and a receiver then listens for a short period of time for the faint sounds that are scattered. The time from the initiation of the sound burst to the receipt of the backscattered sound provides an estimate of the distance to the discontinuity, and the intensity of the return is a measure of the discontinuity's strength. The sounding systems can also detect slight changes in the frequency of the sound, and this Doppler shift is used to measure wind speed. These systems allow wind profiles to be obtained over the lower atmosphere to a maximum height of about 1.5 kilometres.

Observing the Weather

Lightning Detection Networks

Recently, lightning detection systems have been developed that use the electromagnetic wave properties of a lightning discharge to determine the location of the lightning stroke. As is readily apparent when listening to an AM radio during an electrical storm, flashes are accompanied by an immediate static burst on the radio. The strength of the noise (i.e., its amplitude) depends on the proximity of the radio receiver to the discharge. A network of detection stations has been established across North America, each with a detector that consists of an antenna that can determine the direction (or azimuth angle) of a lightning discharge from the antenna. Using the azimuth angles from several of the closest antennae, the discharge's location can be triangulated.

The system in use has a network with detector spacing of 200 to 400 kilometres, which gives it a data collection efficiency of up to 90 percent and a location accuracy of between 8 and 16 kilometres. Lightning stroke information is analyzed by dedicated computing systems, and the location and density of discharges are displayed and updated every hour on a surface map. The movement of areas of lightning strokes and changes in the discharge rates can provide valuable information on the speed and rate of intensification of thunderstorms.

Flying insects stay closer to the ground when the air pressure drops, so when you see swallows flying low to the ground, it could be an indication that rain is on the way.

Communication of Meteorological Data

Availability of appropriate weather data at a central location for analysis, forecast preparation and dissemination all depend on an accessible and reliable communication system. The invention of the telegraph in 1837 presented the first opportunity for meteorologists to undertake weather forecasting programs that could be useful to the public. Dedicated communications networks are used to send data to weather offices within minutes of the collection and analysis of the observation. The internet has made the communication task much less burdensome and has allowed the public to gain access to observations, analyzed data sets and all types of forecasts.

Data Quality Control

Quality control of the observational data is essential to the weather forecast production system. Weather monitoring instruments are maintained and periodically calibrated to ensure that their operation is stable, and that measurement results are repeatable. At one time, trained weather observers and technical staff overviewed the data, but today computers do most of the work. Observations are stored in data banks, where quality control computer programs check for observational range, consistency with previous observations from the site and consistency with nearby observations.

Chapter 10: Forecasting the Weather

Forecasting the weather began long before the science of meteorology was developed. In the early days, large storms were observed and were simply extrapolated downstream in the same direction that storms tended to move. As a result, weather forecasting in Canada started simply as weather observing. An official observatory was established in Toronto in 1839, but it took another 32 years before a national meteorological service was organized in 1871. Observation was the foundation of this service. A late-August 1873 hurricane was the real impetus for the fledgling Meteorological Service of Canada. The hurricane hit the east coast of Cape Breton Island, killing almost 1000 people and destroying 1200 ships and hundreds of homes. Forecasters in Toronto knew that the hurricane was going to hit the island one day before it actually did so, but the telegraph lines were down and no warning could get out. Three years later, storm warnings and general weather forecasts for eastern Canada were instituted. The resulting development of the science of meteorology and environmental prediction, which matured into the current level of understanding, is remarkable, and pace of advancement in meteorological science is comparable to the advancements in flight that occurred during the similar brief time frame.

Forecasting the Weather

> Weather forecasting is not rocket science—forecasting is much more difficult!

Both government organizations and the private sector undertake weather forecasting programs. The government of Canada issues weather forecasts that "are provided for the good of the general population," according to federal policy. Governments worldwide provide a similar service for their populace. Provincial governments are responsible for forecast services that support sectors under their jurisdiction; the Ontario government operates weather forecast programs to support forest fire suppression as well as stream flow and flood warning services. Prediction services operated by the private sector serve the specialized needs that fall outside the government's role. The main focus of this chapter is the federal forecasting process.

Weather forecasts are prepared and disseminated when they will be the most useful for clients. Warnings of hazardous weather conditions are prepared to enhance public safety and reduce property damage. Forecasts are also prepared to support many sectors of the economy, including transportation (aviation and marine transport), recreation (boating, skiing, avalanche warning) and construction. The public wants to know what weather conditions to expect for the day, and the weather forecast is a regular item on newscasts and in the media. In Canada, federal weather forecast centres are located across the country to partition the geographic area into manageable portions. These forecast or storm prediction centres operate similar programs, which include some of the following services:

- preparing public forecasts and bulletins for cities
- issuing warnings for severe weather conditions
- preparing special sector-specific forecasts (transportation, recreation)
- providing information about unusual weather
- providing the latest weather conditions and historical climatic data.

The weather forecast problem is complicated because the atmosphere extends well beyond the local forecasting area. Predictions of the atmosphere at the hemispheric and global scale are prepared at a central prediction centre and are sent to local weather offices. Meteorologists use this guidance material to prepare weather forecasts for several days into the future. In Canada, the central prediction facility is located in Montreal. Although this facility is mandated to support the national weather forecasting program, much of the meteorological information it prepares is also made available to provincial and private forecasting offices.

Numerical Weather Prediction

The atmosphere is a continuous fluid that can be described using a series of mathematical equations that link temperature, pressure, wind, density and moisture. In 1904, Norwegian meteorologist Vilhem Bjerknes recognized that these equations could be used to predict the future state of the atmosphere. In 1922, British meteorologist L.F. Richardson first proposed a computational process to predict the weather using a simplified version of the theoretical equations. He proposed that a short-range forecast could be produced by several groups of individuals using arithmetic calculators to do the many necessary computations. One individual would coordinate the exercise, acting much like an orchestral conductor, ensuring the groups were working together to make the calculations faster than the actual weather was changing. Other individuals would take interim forecasts and make them available to users. Richardson envisioned having a warehouse to store the results for future examination and a research group to develop new forecasting techniques. He actually tried to prepare a six-hour forecast of pressure change using a version of this approach. After much laborious calculation, Richardson's prediction was off by a factor of about 100 because he did not appreciate some of the basic computational techniques that must be used in the process. The massive calculation exercise that was required and the resulting lack of success meant that further exploration of the technique was essentially abandoned for over a quarter of a century. However, Richardson's foresight is now considered quite remarkable.

The full set of equations needed to prepare a weather forecast is too complicated to be solved by simple mathematical techniques, and large computers must be used. The computer-based approach is known as numerical weather prediction (NWP). In the world of theoretical physics, the NWP approach is what is known as an initial-value problem: if the initial state of the atmosphere and the equations that govern its motion are known, its future state can be predicted. The equations must be in a form that allows observed data to be incorporated, which is accomplished through a process called discretization. In the atmospheric NWP model, a three-dimensional grid of small cells is used to represent the atmosphere. Each grid cell has a horizontal length of a few tens of kilometres and a thickness ranging from tens of metres to a few kilometres. A network of several million grid cells is used in the largest and most sophisticated NWP models, which predict the atmosphere over the entire globe.

Forecasting the Weather

Global Weather Prediction Grid

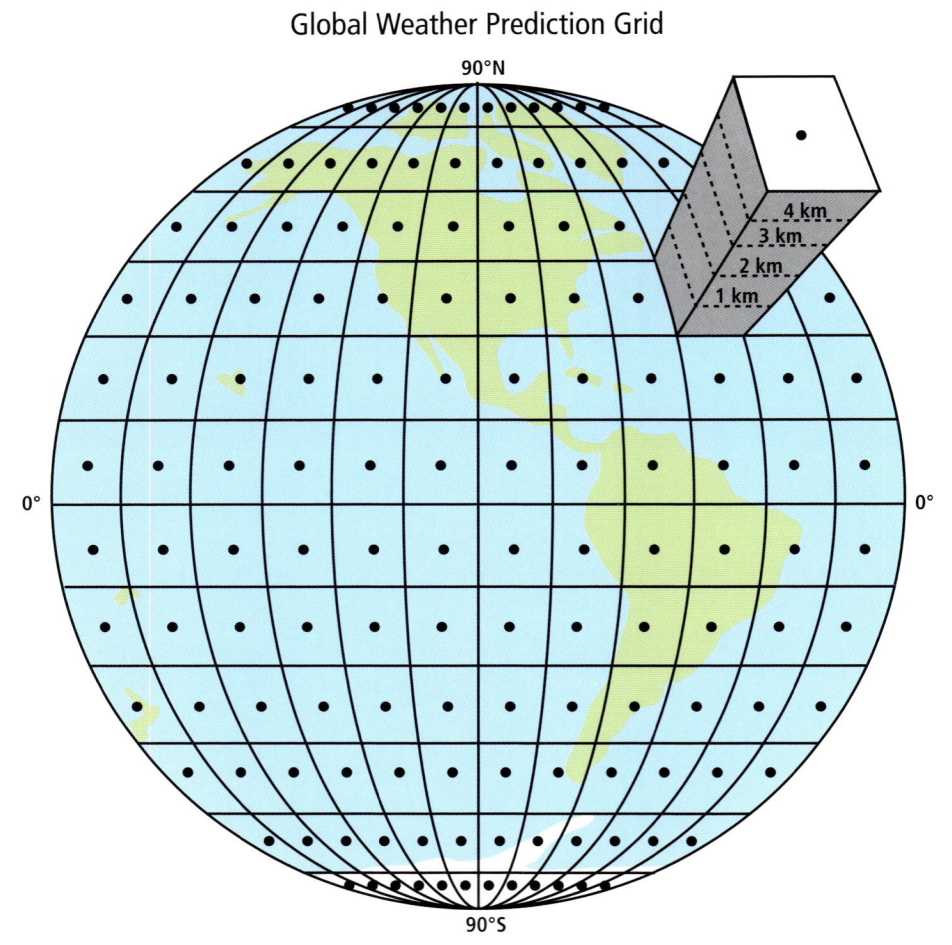

Fig. 10-1 A coarse-mesh grid covering the earth's surface

This three-dimensional grid needs to be initialized with data. Using data from observation stations, temperature, pressure, moisture and wind values are assigned to grid points located at each corner of each grid cell. At points above ground level, temperature, moisture, wind and pressure values are assigned using measurements from radiosonde balloon ascents and from instruments onboard aircraft. Synthetic data generated from weather satellites is used over areas largely devoid of observations, such as the oceans and polar regions. The rate of change of each of the atmospheric parameters is calculated using the equations of motion. At each grid point, a new value of the atmospheric parameter is calculated by multiplying the rate of change by a small time step that is a few minutes long. A short-range forecast of all the parameters is produced, and from this, a new rate

Forecasting the Weather

of change is calculated and another short forecast is computed. This time-stepping process is repeated to generate forecasts out to a few days. The length of each time step is related to the size of the grid cells—the smaller the cells, the smaller the allowable time step.

NWP was only made possible when large, numerical computers became available after World War II. The first successful experiments in NWP were completed at Princeton University in the late 1940s. Although it required almost 24 hours to prepare a one-day forecast using a simple model of the atmosphere, NWP's potential was demonstrated. Since the middle of the 20th century, most national weather predicting programs have used the NWP approach. The NWP computing centres use the fastest supercomputers available, and it still takes several hours to generate all the numerical forecasts. The NWP process is usually repeated twice daily, synchronized with the main 12-hour global weather observation cycle, and the guidance material is sent out to weather forecasting centres.

Numerical Weather Prediction Charts

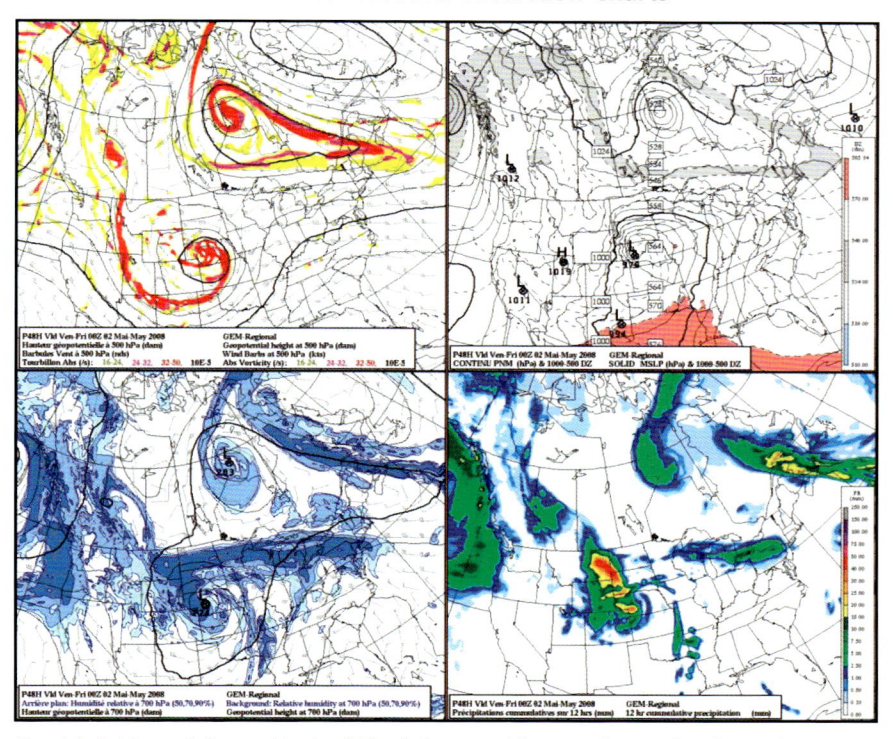

Fig. 10-2 Upper left panel is the 500 mb forecast. Upper right panel is the surface pressure forecast. Lower left is the 700 mb forecast. Lower right is the precipitation forecast. All forecasts are 48 hours.

Forecasting the Weather

The Forecast Products

Warning Program

The cornerstone of the weather forecasting service is the warning program, which issues messages alerting the public of potential weather hazards. Watches are typically issued for larger geographical areas with longer lead times when meteorological conditions are analyzed and diagnosed as being conducive for the development of severe weather events. Watches are intended as a "heads up" that there is the potential for severe weather—during a watch, be alert and make an emergency safety plan. Weather warnings are issued with much more precision for smaller geographical regions and typically have less lead time. A weather warning means that severe weather will soon occur or is occurring—if a warning is issued, it is time to put your safety plan into action.

The rules under which watches and warnings are issued are developed beforehand and are based on the capabilities of the science, public safety considerations and the specific needs of weather sensitive sectors of the economy. Watches are issued up to six hours in advance of a severe summer storm and up to 24 hours in advance of a winter storm. Severe thunderstorm warnings are typically issued two hours in advance of the event. Warnings of severe, large-scale events such as heavy snowfall or wind storms are generally issued at least 12 hours in advance. The forecaster also prepares special weather statements to provide additional explanations of weather conditions that will have a public impact but are not expected to reach the warning or watch criteria (see Figure 8-9 on p. 160 for more information).

Aviation

The Canadian Meteorological Aviation Centre—East in Montreal provides aviation forecasts for Ontario. The aviation industry requires detailed forecasts of many atmospheric parameters, and several different products are prepared. The Terminal Aerodrome Forecast (TAF) is used to enhance aircraft landing and takeoff safety at airports. The TAF includes wind speed and direction, precipitation type and intensity, type of visibility obscurations (fog, precipitation, smoke and dust) and the height of cloud above ground level at the airport. The forecasts are issued every six hours, out to 12 to 24 hours in the future. They are specific about the times that different weather conditions can be expected to change.

Forecasts are also prepared in a graphical form for the area in which aircraft are flying. These maps depict inflight weather conditions across the area and are used by aircraft pilots to infer weather conditions at airports where no TAF is issued. These Graphical Forecast Area charts are issued every six hours for six-hour intervals, out to 18 hours. The maps include the pressure pattern and frontal position forecasts, and they depict areas of cloud, including the expected height of cloud bases and tops, and areas where cloud bases will be near ground level. Areas of precipitation and regions where visibility near the ground

> A lot of people like snow. I find it to be an unnecessary freezing of water.
>
> –Carl Reiner

Forecasting the Weather

might be obscured are also shown. Identifying regions in which aircraft icing conditions might be expected is important for safe aircraft operations, so these areas, including anticipated icing rates, are depicted. Atmospheric turbulence is a concern for passenger comfort and aircraft safety, so areas of turbulence including the degree of severity and levels in the atmosphere where turbulence can be expected are shown. The wind pattern at flight level is useful for aircraft navigation and planning. Maps of winds at several levels in the troposphere are generated from the numerical weather prediction models that provide guidance to the weather forecaster. Again, these products are issued for six-hour intervals, out to 24 hours.

Summer Severe Weather

The summer severe weather program focuses on convective weather and severe thunderstorms. The period of convective concern typically extends from mid-spring until early autumn, though severe convective events have been observed even in December over Ontario. When considering a convective weather forecast, the meteorologist is most concerned about the vertical thermal stability and wind profile of the atmosphere. The development and movement of convective clouds is the forecasting issue. Where will they form, and will they grow explosively to produce heavy rain, hail, wind storms or even tornadoes? The severe weather forecaster needs to have a detailed three-dimensional understanding of the atmosphere. This mental

Forecasting the Weather

picture is built up using radiosonde observations, as well as analyses and predictions of temperature, moisture and wind patterns from the ground up to the tropopause. Images from weather satellites and weather radar provide valuable detail about the structure and motion of the weather systems and existing convective storms. The forecaster must be attuned to the influence of low level moisture from sources such as transpiring forests or croplands. Thunderstorms can develop and grow to severe intensity over a period of an hour or so, and they may move rapidly. The thunderstorms can evolve through various types and modes of convection during their lifetime, and successful meteorologists must anticipate these changes through a thorough analysis and diagnosis of the atmospheric conditions. Watches and warnings for specific sites need to be issued promptly for areas where severe thunderstorms are anticipated so that the public can take safety and security precautions.

Particularly challenging are forecasts of tornadoes, which typically have a lifespan of a few minutes to one hour. Every severe thunderstorm is not a tornado producer. Even sophisticated observational tools, such as the weather radar, do not have high enough resolution to see the small tornadic winds. However, radar can measure wind speed over larger scales using what is known as the Doppler shift technique. Tornadic thunderstorm cells are usually rotating, and the Doppler radar can frequently detect the large-scale rotation of a supercell thunderstorm. The regular observation network is too sparse to delineate small-scale features or observe small-scale events, so the severe weather forecaster depends on observations made by a network of voluntary severe weather watchers. Tornado warnings are issued only when the meteorologist is confident that a tornado is imminent or occurring. The preliminary analysis and diagnosis of the weather situation by the meteorologist is fundamental in anticipating the location and mode of the convection as well as the probability of the various severe convective events. This anticipation allows the meteorologists to continually evaluate new information and to predict severe conditions with sufficient accuracy and advance notice to best safeguard public safety and security. Time is of the essence in issuing the forecast and getting the warning to the public.

Winter Severe Weather

Winter severe weather includes heavy snowfall or rainfall, freezing precipitation, wind chill and strong winds. Winter severe weather usually takes a longer period to develop than does summer severe weather. The large-scale prediction of weather patterns is heavily relied on to determine the location, intensity and motion of the winter storm event. NWP products provide good guidance, but errors in the position of the features can make all the difference in forecasting precipitation rates or precipitation types, and careful attention is paid to pressure patterns and their evolution. Weather observations play a key role, and observations from volunteers are relied on to fill in the details. The start of snowfall is monitored at observation stations, and the forecaster does subjective correlation between snowfall observations and the precipitation patterns displayed by weather radars.

Forecasting the Weather

Temperature and wind predictions are used to prepare wind chill forecasts. Areas of above-freezing temperature layers above the ground help to determine where freezing precipitation may occur. Freezing precipitation is usually difficult to forecast more than a few hours in advance—a problem that is similar to the summer severe weather forecast. Forecasters must anticipate the area where the event may happen and then use all observational data for clues to its occurrence and intensity.

Marine Forecasts

These forecasts are issued during the marine shipping season. In Ontario, the marine forecast is issued for all the Great Lakes except Lake Michigan, which is entirely within the United States. For smaller waterbodies such as Lake Nipissing and Lake Simcoe and for areas along the shores of the Great Lakes, special "nearshore" forecasts are issued to describe the details of the marine weather.

For mariners, the most important parameter is the wind, which can be derived accurately from maps of surface pressure. Waves are built up from the wind blowing for a long time across a water body. Wave height is linked to wind speed, the duration of the wind and what is known as the fetch, which is the length of water over which the wind has blown. Freezing spray is a concern when air temperatures are expected to be below 0° C over fresh water surfaces.

Before modern-day weather forecasting, many people took their cues from nature. For example, when they saw ants moving to higher ground, they knew the air pressure was dropping, so rain could be on the way.

Forecasting the Weather

Phil the Forecaster on nocturnal storm patrol

Because metallic ship superstructures quickly cool to the air temperature, the spray from waves breaking over the ship freezes on contact, and ice can build rapidly. The forecasts issued in late autumn during periods of cold outbreaks include a freezing spray warning. Horizontal visibility is also a concern for safe marine navigation. Fog is the main culprit that affects visibility in marine forecasts. The lake acts as a ready source of moisture, and it can also cool air passing over the surface to below the saturation point, especially in spring and early summer when the water is cold.

The Role of the Forecaster

Meteorologists at the Ontario Storm Prediction Centre in Toronto prepare all public and marine forecasts and warnings for Ontario. Two to four forecasters are on duty every shift, and the team works together to make consistent decisions on the evolution of weather systems across the forecast region. The forecasters in a weather centre coordinate their products with their counterparts in the offices in adjacent regions to ensure that the weather systems are described consistently as they move across the country.

Forecasting the Weather

The weather forecast program operates 24 hours per day every day of the year, so shift work is the reality for forecasting staff. Meteorologists play a vital role in the weather forecasting process. Although observational technology and computerized forecasting tools are sophisticated, the weather forecasting process requires a human touch. The quick synthesis of data and the ability to recognize patterns and make decisions in the face of conflicting or incomplete data are tasks best left to human forecasters.

The first task for a meteorologist starting a shift is to get a briefing from staff members who are finishing a shift. What are the problems of the day? Which weather systems are active in which sections of the forecast region? What forecasts and warnings have been issued? What is the confidence level that conditions will evolve as predicted? Today's weather office is moving toward a paperless working environment, and the computer-supported graphical workstation has replaced the map displays that were once the predominant feature. The meteorologist reviews observed conditions over the region of forecast responsibility and often undertakes a hand analysis of the computer-plotted surface weather map. This hand analysis is one of the few paper products still used because the process helps focus the meteorologist's attention on the details and occasional data anomalies and inconsistencies. Hourly weather data from automated and staffed observation sites across the forecast region is continuously streaming into the office through the communications system. Computers organize the data, and computer algorithms compare the observations with the valid forecast. If there is a significant discrepancy, the meteorologist is notified so that an amendment can be issued.

A large variety of other depictions of the atmosphere are generated. Weather patterns at various levels in the vertical, and graphs of the vertical temperature and moisture structure provide the meteorologist with knowledge of atmospheric conditions favourable for convection. Imagery from geostationary and polar-orbiting satellites is continuously displayed. Displays of weather radar images provide a time sequence of radar patterns and precipitation rates across the province, and the lightning detection system indicates where thunderstorms are active. The meteorologists use their knowledge of atmospheric processes and the influence of topographic and other controlling features to diagnose the current atmospheric condition and its short-term change.

The forecast products are all prepared at the computer workstation. The computer automatically generates a time sequence of all the weather elements, such as cloud, temperature, wind, precipitation and visibility, to be included in the forecast for all locations in the region. The meteorologist makes any necessary adjustments to the weather element sequence using an interactive knowledge-based system. When the forecaster is satisfied with the forecast, a command is sent to the computer system to generate the worded forecast, which is transmitted to media outlets. The product is also sent to a voice generation system that automatically prepares an audio clip for broadcast on the weather radio network across the province.

Forecasting the Weather

For aviation, Terminal Aerodrome Forecasts (TAFs) are prepared for each of the larger airports across the province. Again, the computer guidance material is used, but the coded forecast is input directly into a bulletin by the meteorologist. Aviation forecasts include many weather parameters, and a continuous comparison is made between observations and forecasts. The comparison is largely performed by the computing system, which automatically sends out an alert to the forecaster if discrepancies between the forecast and observed values exceed predefined limits. The meteorologist must quickly evaluate the reasons for the discrepancy and issue an amendment to the TAF.

The forecaster also composes prognostic Graphical Forecast Area charts directly on an interactive aviation workstation. Again, the material generated by the central computer is used as a guide. The forecast surface pressure patterns are based on the NWP guidance. The forecaster adds frontal positions and outlines areas of cloud and precipitation, as well as regions where icing and turbulence are to be expected. All these fields are developed from the temperature, moisture and wind patterns in the NWP guidance.

Old weather saying:
As high as the weeds grow, so will the bank of snow.

These common cloud edges are "deformation zones"—the leading edges of moisture associated with an advancing low-pressure area. Observing these and other cloud features allows the informed weather watcher to read the sky like a book. One just has to learn the vocabulary of the sky (clouds).

Chapter 11: Air Quality

Ontario has a diversified economy and a population that currently comprises one-third of Canada's total and is continually growing. Although population growth in Ontario's cities is more rapid than in the rural areas (80 percent of Ontarians live in the urban areas along the shores of the Great Lakes), both homegrown and imported air pollution span the province. Up to 50 percent of the pollution over southwestern Ontario originates in the United States.

An air quality advisory is issued by Environment Canada when air quality is expected to be poor (the Air Quality Index is expected to reach or exceed 50). These advisories are issued in cooperation with the Ontario Ministry of Environment, the operator of the Air Quality Index Program.

Air Quality

Emissions to the Atmosphere

The principal cause of air pollution is human activity. Transportation is the number one offender, followed by several industrial sources, including fossil fuel-fired power generation, iron and steel, cement and concrete manufacturing, petroleum refining, pulp and paper, base-metal smelting and chemical processing. Residential wood stoves are also a significant source of air pollution in the form of particulate matter in Ontario.

Air quality is often worse near the shores of the Great Lakes because breezes off the lakes concentrate pollutants a short distance inland. Through continued monitoring programs, Environment Canada scientists are working to better understand the role that the Great Lakes play in Ontario's air quality.

Pollutants are classed as either primary or secondary. Primary pollutants have well-defined sources such as smokestacks and vehicle tailpipes. Secondary pollutants are those that are the product of chemical reactions in the atmosphere. These reactions occur among the primary pollutants and with various constituent gases in the clean atmosphere.

High temperature combustion of fuels produces most of the emissions of concern. Nitrogen oxides are produced when any combustion occurs in the oxygen/nitrogen gas mix of the atmosphere. They may also be produced by the oxidation of

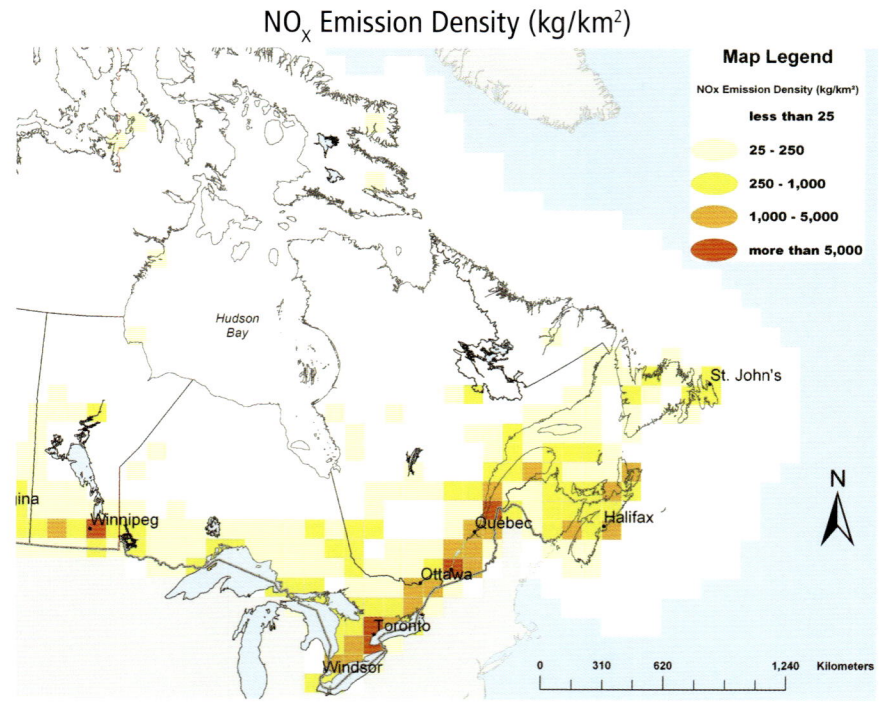

NO_x Emission Density (kg/km^2)

Map Legend

NOx Emission Density (kg/km²)
- less than 25
- 25 - 250
- 250 - 1,000
- 1,000 - 5,000
- more than 5,000

Fig. 11-1

Air Quality

nitrogenous compounds that are in the fuel. Nitric oxide (NO) and nitrogen dioxide (NO_2)—collectively known as NO_X—and nitrous oxide (N_2O) are the most important nitrogen pollutants. Nitrous oxide is also significant because of its radiative properties. It strongly absorbs infrared radiation and is one of the main greenhouse gases. Ontario is responsible for 21 percent of Canada's 2427 kilotonnes of NO_X emissions. Mobile transportation sources (vehicles and aircraft) account for 65 percent of Ontario's emissions, with the rest coming from stationary sources (power plants and residential and industrial heating). Carbon monoxide (CO) and carbon dioxide (CO_2) are also products of combustion and are formed when the carbon in fossil fuels is oxidized by atmospheric oxygen.

Many fossil fuels contain trace amounts of sulphur and, though much of the sulphur is extracted from fuels, residual amounts are converted to sulphur dioxide (SO_2) during combustion. Hydrogen sulphide (H_2S) is naturally present in natural gas and oil reservoirs. It is a very toxic gas with a characteristic rotten egg smell. In the production of fossil fuels, H_2S is extracted and converted to elemental sulphur, but small amounts of residual H_2S are released or transformed into SO_2 through combustion. Ontario produces about 24 percent of Canada's SO_2 emissions; almost half of this total is the result of non-ferrous mining and smelting. Ontario has committed to reducing its emissions by 50 percent by 2015.

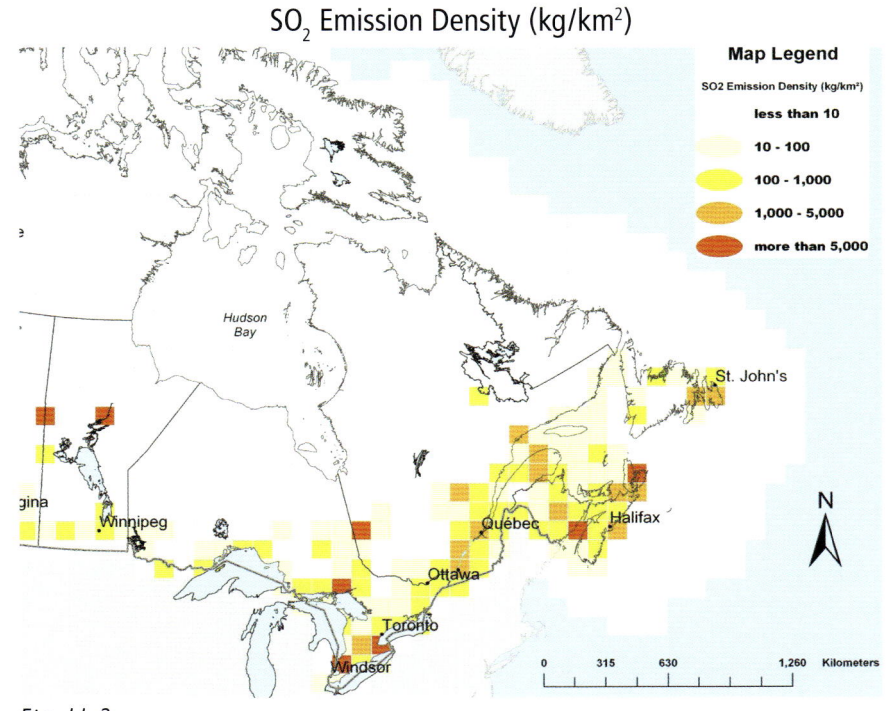

SO_2 Emission Density (kg/km²)

Fig. 11-2

Air Quality

A wide variety of hydrocarbons find their way into the atmosphere. These volatile organic compounds (VOCs) originate from the evaporation and combustion of liquid fossil fuels and from chemical manufacturing, petroleum production and solvent use in chemical processes and domestic applications. Ontario contributes about 23 percent of Canada's VOC emissions.

Ammonia (NH_3) plays a role in the production of particulate matter (PM) in the atmosphere. Animal husbandry, fertilizer application and some industrial processes are sources of ammonia in Ontario. Ontario contributes about 27 percent of Canada's ammonia emissions.

Atmospheric particulates are contaminants that are attracting increasing concern. Some particulates are classed as primary pollutants, but many particulates are secondary pollutants that form in the atmosphere. Particulate matter impairs the clarity of the atmosphere by degrading its light transmission properties, resulting in reduced visibility of distant objects. It can also have negative health impacts. Particulates are generally categorized according to their aerodynamic size based on their ability to be deposited in lungs. Larger particles usually have a natural origin, such as windblown dust from dry soil or re-suspended dust from road surfaces, but they can also originate from combustion as smoke, soot and ash. These particles are generally larger than 10 microns in diameter. It is the smaller particles—generally with a size less than about 2.5 microns—that find their way into the lungs and cause the greatest health risk. These particles may be either solid or liquid droplets and are often formed from the gases that are a product of combustion. Ontario contributes about 23 percent of Canada's 2.5 micron particulate matter (PM2.5) emissions.

Compounds in the categories of toxics, heavy metals and persistent organic pollutants are either minor by-products of industrial processes or are heavily regulated, and few reach the atmosphere in Ontario.

Emissions of pollutants vary greatly across Ontario. With their high population concentrations and intense levels of industrial activity, urban centres and their surrounding areas emit the majority of pollutants. Agricultural emissions are widespread in the cultivated lands in the southern half of the province. Industries report their annual emissions to the Canadian government for archiving in the National Pollution Release Inventory. Ontario has more than twice as many sources in the inventory than any other Canadian jurisdiction.

There is an atmospheric cycle through which pollutants are mixed, moved around, sometimes transformed and finally removed from the atmosphere by several natural processes. Polluting gases are diluted downwind of the emission sources. Although the saying "the solution to pollution is dilution" was often heard, dilution is taxing on the clean air resource. Consider vehicle emissions of carbon monoxide. In 1965, the average emission rate was 30 grams of CO per kilometre of travel. About 2 million litres of clean air—the amount breathed daily by 200 people—are required to dilute that much CO to levels that are safe for humans to breathe. So, it takes large volumes of air to dilute all the pollutants to levels that are considered safe.

Air Quality

Emissions Charts for Ontario

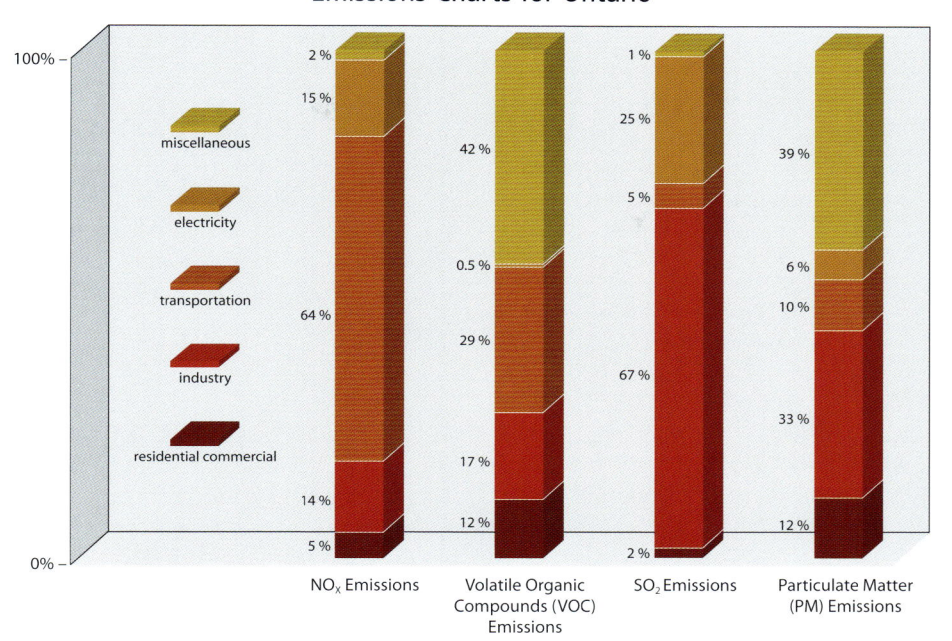

- miscellaneous
- electricity
- transportation
- industry
- residential commercial

NO$_x$ Emissions: 2%, 15%, 64%, 14%, 5%

Volatile Organic Compounds (VOC) Emissions: 42%, 0.5%, 29%, 17%, 12%

SO$_2$ Emissions: 1%, 25%, 5%, 67%, 2%

Particulate Matter (PM) Emissions: 39%, 6%, 10%, 33%, 12%

Emissions Charts for Ontario

NO$_x$ Emission Sources Canada (2005):
- transportation 53%
- upstream oil and gas 19%
- electric power generation 10%
- other sources 18%

Greenhouse Gas (GHG) Emissions:
- waste 4%
- agro-ecosystems 6%
- transportation (vehicles) 27%
- fossil fuel industries 4%
- commercial/ public admin. 6%
- residential and agriculture 13%
- total industrial 29%
- power generation 11%

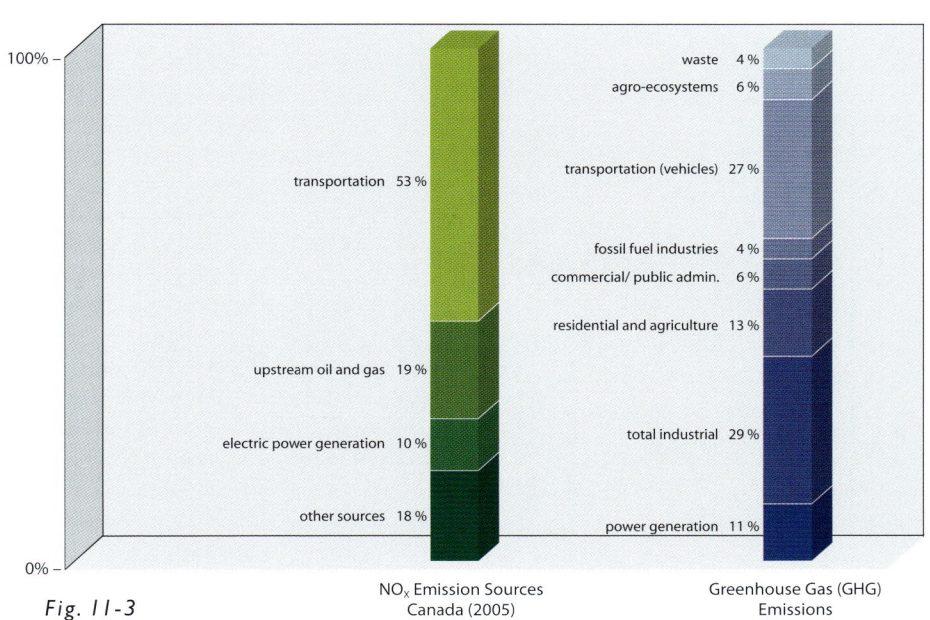

Fig. 11-3

197

Air Quality

In many regions of the world, there is just not enough clean air, and an emphasis was placed on improving vehicle efficiency along with reducing pollutant emission rates. By 1997, the emission standard for CO was reduced to 2.1 grams per kilometre. On a per vehicle basis, we are now polluting significantly less air.

Cloud condensation nuclei are in part composed of the small particles originating from pollutants. Studies of rainfall patterns have revealed how pollution in the atmosphere influences precipitation. For example, enhanced rainfall within Chicago relative to its less polluted suburbs has been attributed to the vast number of condensation nuclei injected into the atmosphere by industrialized regions in that city. Farther downwind, enhanced rainfall in the city of LaPorte, Indiana, has been attributed to the transport of nuclei from the same upwind Chicago pollution sources. The rainfall process is a cleansing mechanism for polluted atmospheres—it removes and washes away some nuclei.

Air pollution and smog are major problems in Ontario. The province's residents are exposed to the most days with poor air quality in the country and some of the worst air quality overall. The impacts of this pollution can be most significant in the rural areas downwind of the source urban areas.

Pollution Movement in the Planetary Boundary Layer

The lowest layer of the atmosphere is known as the planetary boundary layer (PBL). Within the PBL, most life forms are subjected to air pollution. Air quality and the dispersion of pollutants are influenced by the turbulent motion and temperature structure within the PBL. Three key factors determine the characteristics of the PBL:

i) transport of heat from the earth's surface. Heating caused by the solar radiation cycle, heating from phase changes of water (latent heat) and the spatial variation in heating (near lakes and over different land surfaces) are all examples of the complex heating process. Heat is an important energy source that leads to mixing of the PBL—just like the heating element on a stove causes water in a pot to turn over and mix.

ii) influence of roughness at the earth's surface. Airflow is disturbed by the presence of objects, from blades of grass or trees to buildings, hills and mountains. These objects result in a surface roughness that slows the airflow in the PBL by imposing what is known as frictional drag. Large objects can also deflect and block airflow. Roughness features induce mechanical turbulence, much like water flowing over a rough riverbed is turbulent. The larger the roughness features and the stronger the speed of the flow, the greater the degree of turbulence.

iii) airflow and thermal structure of the atmosphere at the top of the PBL.

Air Quality

This level is the bottom of what is known as the free atmosphere and is characterized by the atmospheric flow associated with large-scale pressure patterns. At this level, the geostrophic wind law describes the horizontal flow. Upward and downward motions associated with the large-scale flow are also evident. The top of the PBL can be a sharp transition zone into the free atmosphere. If the temperature in the PBL is lower than in the immediately adjacent free atmosphere, the top of the PBL effectively acts as a lid that can prevent the upward motion of pollutants out of the PBL.

Turbulence

Atmospheric motion is anything but uniform—it is turbulent. Whether winds are light or strong, we sense the turbulence as gustiness. When wind measurements are recorded, the time series shows frequent changes in wind speed and direction. These fluctuations are characteristic of turbulent flow. Air quality scientists think of wind as having two components. The first component we can call wind flow. It is considered to be steady for some interval of time. The second component is perturbation, which is the slight departure from the mean wind. The motion of air pollutants is governed by the wind flow itself and by perturbations in that flow. Wind flow is a vector quantity—it has speed and direction, and it is governed by the mechanical equations of motion of the atmospheric fluid. The mathematics dealing with perturbations in the wind, known as the turbulent component of the airflow, are complex. An old saying that demonstrates the difficulty of understanding turbulence goes, "When you get

Depiction of a Plume Rising in a Stable Atmosphere

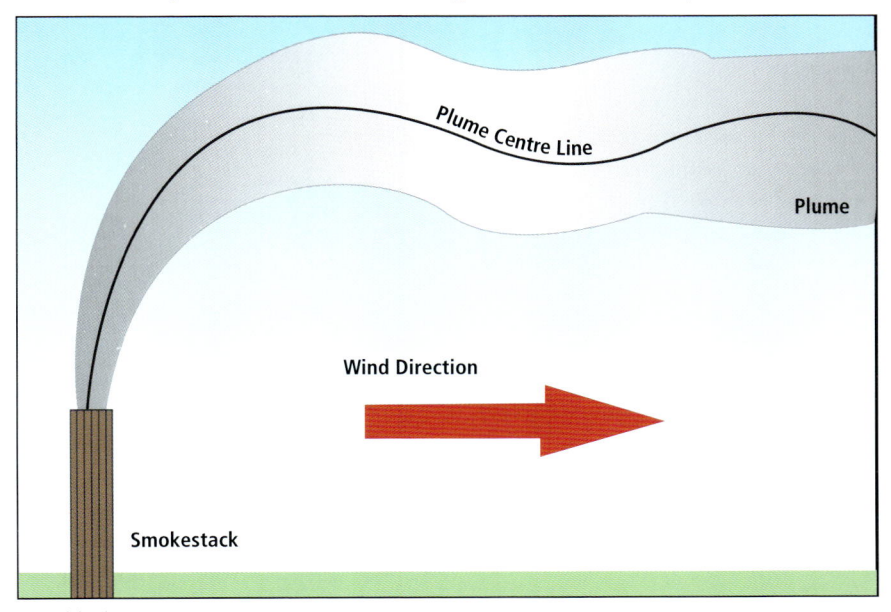

Fig. 11-4

Air Quality

to heaven, God will offer two explanations—one for quantum physics, and the second for turbulence theory." Air quality science focuses on the effects of such turbulent motion.

The rate at which pollutants disperse and how pollution concentrations change with time is described by the mathematics of turbulence theory. This theory can be put in the form of mathematical equations, which are used to calculate the rate of spread of pollutants and the changes in pollutant concentration downwind of an emission source.

When pollutants are emitted into the atmosphere from a point source, such as a smokestack, they have two key characteristics. If they are warmer than the surrounding air, their buoyancy causes the pollutants to rise. They also immediately start mixing because of turbulence and spread out laterally from the source. If a wind is associated with the fluid, the pollutant cloud takes the form of a plume. The plume has a characteristic shape, as shown in Figure 11-4. It initially moves vertically then is bent over by the wind. With time, the polluting gas spreads away from the centre line, taking on a conical form. The shape of the cone, and the pollutant concentration within the cone, is described by the statistical properties of the turbulent atmosphere.

> The Windsor–Sarnia corridor generally has the highest multi-pollutant concentrations, when you consider ozone, PM2.5, NO_2 and SO_2.

Temperature Structure

The temperature structure of the planetary boundary layer has an important influence on air quality because the vertical temperature structure controls the vertical mixing. As mentioned earlier, the rate of change of temperature with height is known as the lapse rate. The lapse rate is usually negative—that is, temperatures decrease with increasing height. The greater the rate of temperature decrease with height, the greater the ability of the atmosphere to disperse pollutants. Most pollutants are emitted near the ground, so the lapse rate in the PBL is important in dispersing pollutants. However, there are many situations in which temperatures increase with height, creating an inversion. An inversion condition in the atmosphere slows the dispersion process, and pollutant concentrations can increase near a source that is continuously emitting.

The lapse rate varies widely during a 24-hour period. When the sun sets, the ground surface cools, as does the air in contact with the ground. Air farther aloft does not cool as quickly, and a nocturnal inversion develops. Pollutants emitted into this inversion layer may move laterally with the wind, but they have little vertical movement. The nocturnal inversion often traps pollutants near the ground.

Other types of inversions are also important in Ontario. The interface between two different air masses is known as a front. Frontal inversions can play a role in trapping pollutants near the ground. The frontal surface can have a shallow slope, and a large area of the province can be subjected to frontal

Air Quality

inversion effects, especially in the cold seasons with slowly moving warm fronts. Cold fronts, on the other hand, are often rapidly moving, and they clean out stagnant polluted air masses. Cooler, drier and typically much cleaner air from the northwest follows Ontario cold fronts.

Terrain can also play a major role. Cooler air pools in river valleys and may be overrun by warmer air. The resulting inversion can trap pollutants within a valley. This phenomenon can be seen in the Ottawa Valley, especially when a high pressure area in Québec directs cold, easterly surface winds up the valley of the Ottawa River and the St. Lawrence.

In December 1952, more than 4000 people died in London, England, as a result of a severe pollution event. A strong thermal inversion acted as a lid and trapped moisture and pollutants in the PBL above the city. Incident sunlight was reflected by higher-level cloud decks and by the top of the fog layer itself. Sulphur-laden smoke from coal-fired furnaces and stoves accumulated over a period of five days as the inversion persisted. Most people died because of respiratory tract infections, such as bronchitis and bronchopneumonia. Largely because of this event, the term smog—meaning a mixture of smoke and fog—was coined.

London, England, used to be known for its so-called pea-soupers: heavy blankets of smog, made up of a combination of fog and smoke from factory smokestacks and the burning of coal.

Air Quality

Chemical Transformation in the Atmosphere

Photochemical smog is common in all large cities, where vehicles, industries and other pollution sources are concentrated. Most prevalent and well known were the smog events that frequented greater Los Angeles. Through the 1960s and 70s, the causative chemicals and chemical reactions responsible for the smog were determined, and a solution to the issue has progressed with California's imposition of stringent vehicle emission controls. Photochemical smog is now a much less frequent occurrence, and events are less severe than during its peak in the 1970s.

Photochemical smog consists of a mixture of the primary emissions of NO_X and VOC gases, and the secondary pollutants ozone (O_3) and particulates. These two secondary pollutants are produced when intense sunlight acts on the mixture of primary emission gases.

Reactions in the polluted mixture proceed depending on the concentrations of the precursor gases, the temperature and the amount of mixing that is present in the planetary boundary layer. Some of the reactions are rapid, some are slow and some depend on the intensity of sunlight. Generally, the reactions in the polluted soup proceed to what is known as an equilibrium state. In the worst-case scenario, this equilibrium state has high concentrations of ozone and particulates.

Periods of high ozone can last several days and frequently occur when a stagnant air mass traps pollutants over a region. Recent studies have shown that every major Canadian urban centre has levels of ground-level ozone high enough to pose a health risk. Ozone is a powerful greenhouse gas that contributes to climate change and is also known to damage vegetation and cause the deterioration of some natural and synthetic materials, including paints and dyes.

Formation of Photochemical Smog

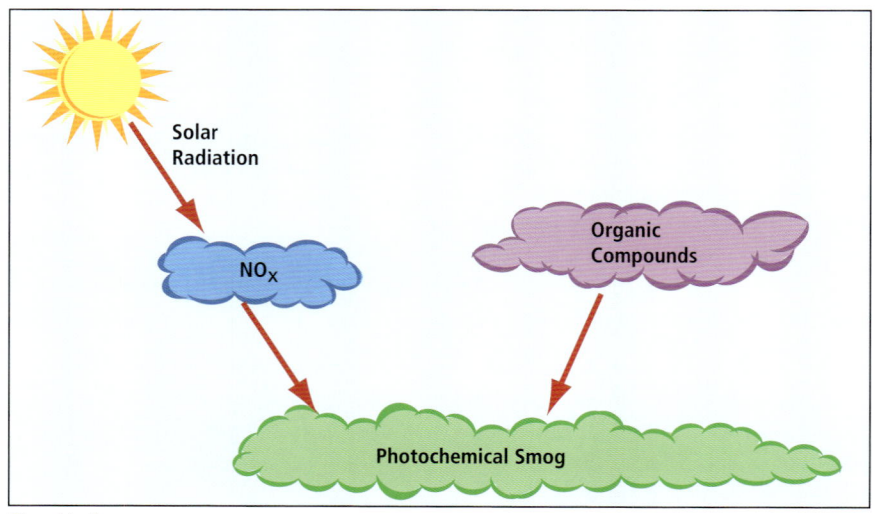

Fig. 11-5

Air Quality

Ozone Map

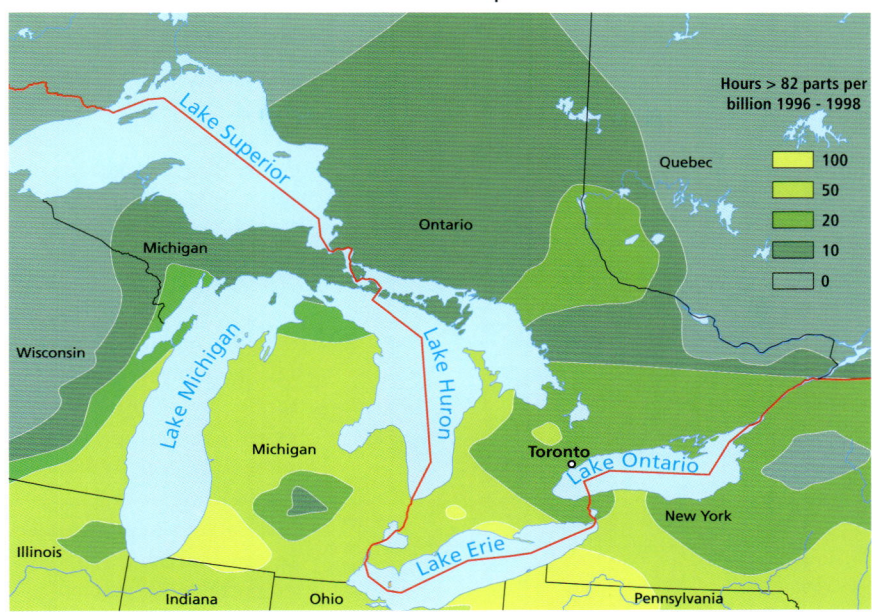

Fig. 11-6

Ontario's major cities have significant emissions of the pollutants that can lead to the production of photochemical smog. The presence of NO_2 is evident because this gas preferentially absorbs light in the short wavelengths (blue to green). When light passes through the gas, these short wavelengths are absorbed, allowing the longer wavelengths—the red to orange colours—to pass through the gas. As a result, the gas appears to have a brownish colour. The brownish tinge is usually evident downwind of Windsor, Sarnia and Toronto during light wind conditions in summer and winter. These cities can also have meteorological conditions in summer that favour the production of photochemical smog (calm winds and intense sunlight). Because the reactions can take a few hours, peaks in concentrations of ozone and small particulates vary in space and with time. Air quality monitors near downtown Windsor and Toronto usually show elevated NO_X concentrations, especially during the morning rush hour when there is more NO_X available than is necessary for ozone production. During a typical day, NO_X levels peak during morning and afternoon rush hours. Ozone reaches its highest level late in the afternoon when solar intensity reaches a peak downwind of the city centre and the photochemical conversion process is most active. Overnight, concentration levels of most of the primary and secondary gases drop to a minimum.

Smog situations can also occur in winter. They characteristically have high concentrations of PM and NO_X, but ozone levels are low because of low temperatures

Air Quality

5-year mean precipitation pH

pH
4.2
4.4
4.6
4.8
5.0
5.2
5.4

5 Year Mean Precipitation pH / Précipitation de pH moyenne sur 5 ans

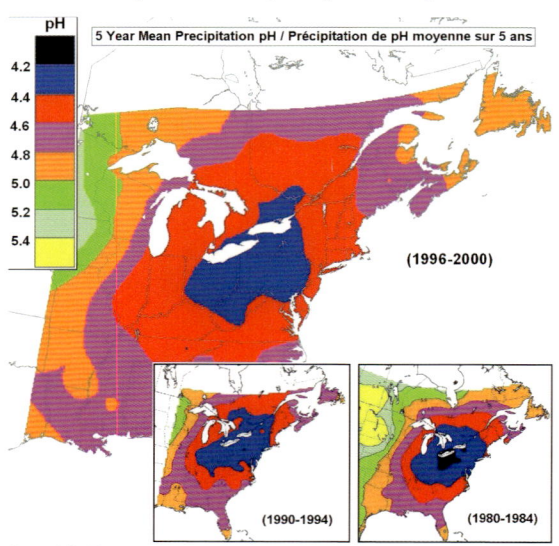

(1996-2000)

(1990-1994)   (1980-1984)

Fig. 11-7

and weakened solar intensity. Winter smog usually forms because of the presence of strong thermal inversions and low wind speeds that trap pollutants near the ground. Chemical reactions proceed slowly at low temperatures, though low temperature conditions are favourable for the formation of some types of particulates.

The formation of acids in the atmosphere and the deposition of these acids on lakes and forests is an issue in much of eastern North America and Europe. Sulphur dioxide and NO_X both play a role, and the high rates of emissions of these gases in Ontario have been the focus of a considerable research effort. Sulphur dioxide has been the primary culprit in the acid rain issue. It transforms into sulphuric acid in the presence of water vapour and other atmospheric chemicals known as free radicals. Nitric acid (HNO_3) forms in the atmosphere as a byproduct of the same chemical reactions involving NO and NO_2 that are responsible for ozone production. The gaseous forms of the acids can be deposited in dry form, and the acids in aqueous form are deposited in precipitation known as acid rain.

Negative impacts of acid deposition usually depend on the nature of the receptor. Acids can be neutralized or rendered less harmful by some chemicals that occur naturally in many receptors. Most of the receptor surfaces in Ontario have this neutralizing capability, known as a high buffering capacity. However, the groundcover and lakes in areas of Precambrian rock (the areas of central and northern Ontario associated with the Canadian Shield) have low buffering capacity and are susceptible to acidification. As well, because the acids and the chemicals that are acid precursors can be transported long distances by the winds, acid deposition can occur far from the pollution source. This chemical transport is the reason for much of the acidification of sensitive lakes in eastern Canada and the United States. Many of the polluting sources are well upwind, and long-range transport of sulphur compounds is the primary culprit. The above map displays the level of acidification in precipitation as a result of the ingestion of the noted atmospheric pollutants. The map suggests some improvement has been made in tackling the acid rain problem.

Air Quality

Forest fires can also have a major influence on air quality in Ontario. The smoke, of course, has a high particulate loading and can be responsible for health distress. Reports of associated illness and increased admissions to medical centres occur at locations well removed from the fires. Chemical reactions also occur in the smoke plume. Many of the reactions and chemicals are the same ones involved in the photochemical smog problem. Smoke plumes from forest fires contain elevated levels of NO_2 and VOCs produced from the burning of massive quantities of biomass. Some of the highest levels of ozone and particulate matter have been measured at sites beneath smoke from forest fires when ozone and small particulates are mixed down to ground level.

High ozone concentrations associated with smoke plumes from distant forest fires have been measured in Ontario, and elevated pollution levels in the southeastern United States have been attributed to forest fires in northwestern Canada.

Smoke from forest fires burning along the Mackenzie River in July 2008 reached southern Ontario and the eastern United States. The skies were remarkably hazy, and the sunrises and sunsets were very colourful.

On July 2 and 3, 2002, thunderstorms sparked more than 85 fires in central and western Québec. These fires intensified on July 5, and by July 8, about 250,000 acres of forest had been burnt. The smoke from the fanned fires was transported rapidly and with little dispersion into southern Québec, central and eastern Ontario and the northeastern U.S. because of a low pressure system over Nova Scotia. The smoke was so thick that people thought their homes were on fire. The accompanying satellite image shows the fires (in red) and the long, grey smoke plume directed southward.

Air Quality as a Societal Issue in Ontario

Ontarians are concerned with maintaining a healthy environment. Governments usually assume the lead in establishing criteria to protect air quality and human health, and Ontario has established air quality objectives for many air contaminants that are meshed with national goals. The Ontario ambient air quality objectives are expressed in terms of pollutant concentration levels in the atmosphere. Averaging times for pollutant measurements are specified, usually as hourly averages, though daily and annual averages are also used. These averaging times are closely linked to impacts—for example, how much time a plant, animal or human will be exposed to the pollutant. The numerical values are established to protect the environment as well as human health.

Air Quality

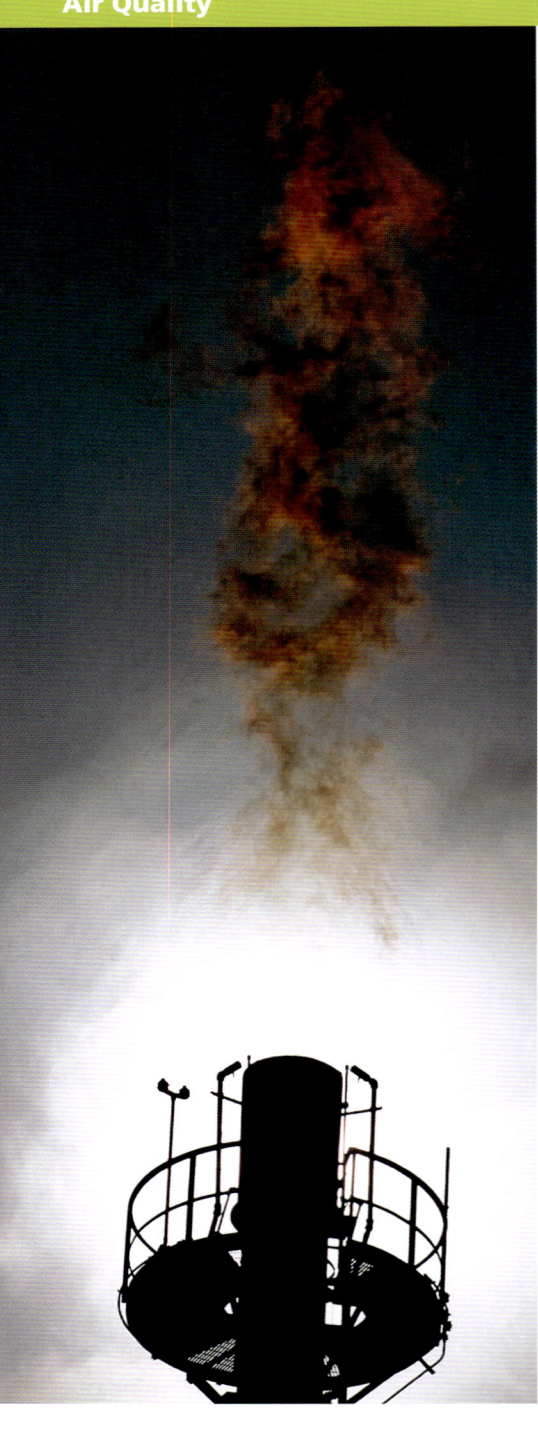

In addition to ambient air quality objectives, governments also establish regulations on emissions from various sources of pollutants. Ontario's Drive Clean Program is an emissions inspection and maintenance program for vehicles that has the goal of reducing vehicle emissions by 22 percent in the program area, which spans from Sarnia to Peterborough. The federal government has also enacted regulations on emissions from cars and trucks aimed at reducing the levels of NO_X in cities. Various laws and programs have been aimed at reducing the times that vehicles are allowed to idle. At the same time, the provincial government has enacted regulations to reduce emissions from sources such as coal-fired power plants and from flares operated by the oil and gas industry.

In Ontario, the government, industry and the public all have a role to play in dealing with the air quality issue. Several processes are in place to ensure that the three groups of stakeholders undertake joint discussions on many aspects of air quality. Areas for consideration include the setting of air quality objectives, the management of local air quality concerns, and approval of applications for new facilities that may have sources of air pollutants. The groups also evaluate the state of scientific knowledge on many aspects of provincial air quality, and if it is deemed insufficient, they may propose that detailed investigations and research be undertaken.

Chapter 12: The Changing Atmosphere

The atmosphere includes a number of gases that have concentrations at trace levels. Some of the gases have a short residence time in the atmosphere, but some are long lived. Many of these trace gases play an important role because of their physical characteristics or because of their ability to participate in chemical reactions. These gases are important to the earth's environment and, indeed, to the health of the ecosystem. One of the more significant groups of these gases is known as the radiatively active gases, so-called because of the role they play in interacting with electromagnetic radiation—either the radiation emitted by the sun or by the earth itself. Some of the gases have a natural origin, and some are only present because people release them into the atmosphere. Ozone and the greenhouse gases play a particularly important role for our planet.

> *Evidence of climate change can be seen in ecosystem changes— for example, the Virginia opossum now flourishes in central and southwestern Ontario.*

The Changing Atmosphere

Formation of Stratospheric Ozone

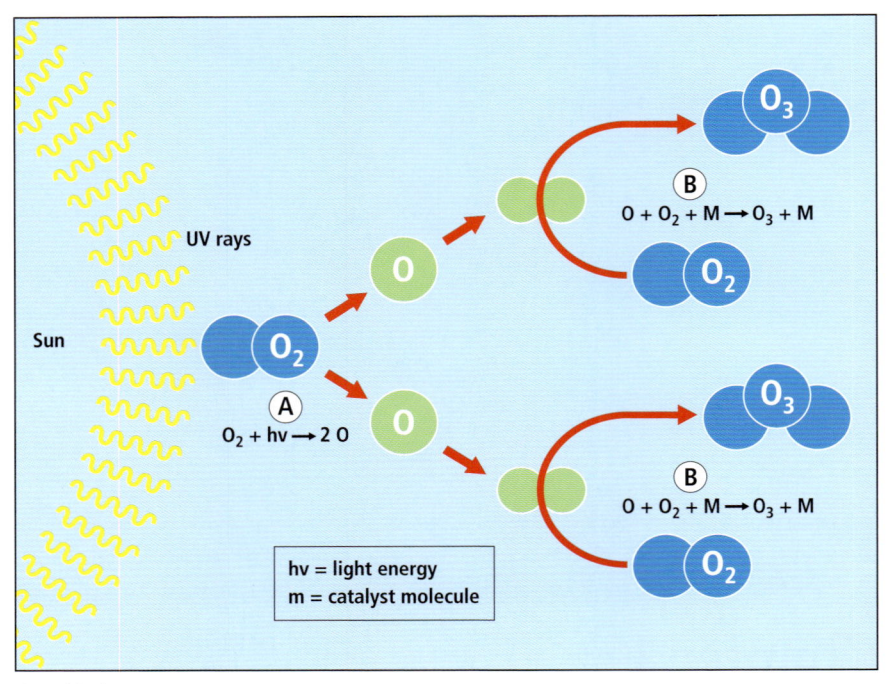

A — $O_2 + h\nu \rightarrow 2O$
B — $O + O_2 + M \rightarrow O_3 + M$

$h\nu$ = light energy
m = catalyst molecule

Fig. 12-1
A - ultraviolet light strikes oxygen, forming atomic oxygen
B - ozone molecules form in the presence of a catalyst molecule

Stratospheric Ozone

Ozone is a form of oxygen gas that has a dual role in the atmosphere—it is in some ways beneficial and in some ways detrimental. Ozone gas contains three oxygen atoms and is present at low concentrations throughout the atmospheric vertical column from the surface to the stratosphere. Ozone in the stratosphere is produced through natural processes. The ozone molecules absorb much of the short-wave ultraviolet radiation emitted by the sun, protecting life forms from the damaging effects of ultraviolet solar radiation. This property of ozone has been known since the 19th century. Some ozone in the lower atmosphere is produced through the chemical reactions between gases that have natural and anthropogenic origins. This low-level ozone is detrimental to human health and the ecosystem, essentially because ozone is reactive with other substances. Ozone also plays a small role as a greenhouse gas. In the lower atmosphere, ozone contributes to warming of the atmosphere, but stratospheric ozone has a cooling effect. The phrase "good up high, bad nearby" describes the dual role of ozone.

The Changing Atmosphere

Formation of Stratospheric Ozone

In the upper atmosphere, ultraviolet solar radiation at short wavelengths causes the dissociation of some oxygen molecules into atomic oxygen. The atomic oxygen reacts quickly with oxygen molecules to form the ozone molecule. This reaction also involves a third gas, usually the ever-abundant nitrogen. Nitrogen itself is unaffected, except that it absorbs some of the energy released in the formation of ozone.

The ozone molecule, in turn, absorbs ultraviolet solar radiation and splits apart to form diatomic (O_2) and atomic (O) oxygen. This cycle, in which ultraviolet light interacts with oxygen and ozone, is in natural balance in the mid-stratosphere and effectively absorbs harmful short-wave ultraviolet radiation. The intensity of solar radiation is greatest at the equator, and that is where most of the ozone is produced. This ozone is transported toward the polar regions by the large-scale, or Hadley, atmospheric circulation pattern (see pp. 16–17).

However, substances emitted into the atmosphere are upsetting the natural balance. The gases of concern are the nitrogen oxides, formed in the combustion process, and other gases including the chlorofluorocarbons (CFCs), which contain the chlorine atom, and chemicals containing the bromine atom. Nitrogen oxides (NO_X) emitted by combustion in the lower atmosphere normally undergo chemical reactions fairly quickly and do not find their way into the stratosphere, so the concern is with NO_X emitted at high altitudes. Although there were concerns over NO_X emissions from the fleet of supersonic aircraft proposed in the 1970s, there are relatively few aircraft flying at these critical heights, and the resultant ozone depletion from that source has been relatively small to date.

Of much greater concern is the role of CFCs and bromine compounds. When CFCs were first discovered in the early 20th century, they were much valued because they are stable and don't react with other substances. CFCs were used as propellants in aerosol spray cans, and they replaced the dangerous ammonia gas in most home refrigeration systems. Chemicals containing bromine include methyl bromide, used as an agricultural fumigant, and the halons that are used in fire retardants. The problem with these gases is that they have a long lifespan. When introduced into the lower troposphere, these gases eventually work their way into the upper atmosphere through the action of thunderstorms, which are frequently so vigorous that their tops push into the lower stratosphere. The most energetic thunderstorms are at equatorial latitudes, so most of the penetration into the stratosphere takes place there. These long-lived gases are transported farther into the mid-stratosphere and eventually toward the polar regions by the same Hadley circulation that moves ozone toward the poles.

High energy ultraviolet solar radiation that enters the stratosphere breaks the chemical bonds in the stable compounds, releasing chlorine and bromine atoms into the stratospheric ozone layer. These atoms undergo reactions, forming what are known as catalyst molecules. The catalyst molecules react with the ozone molecules, breaking the ozone apart while the catalyst remains intact. The

The Changing Atmosphere

Increased levels of UV radiation can expose us to a greater chance of skin damage from sunburn.

catalyst goes on to break apart another ozone molecule, and so on. Each chlorine or bromine catalyst molecule can effectively destroy thousands of ozone molecules.

The catalyst reactions that dissociate ozone do not proceed at a uniform rate or equally across the stratosphere. The most pronounced effects are in the polar regions where, during the cold season, stratospheric temperatures can drop to below -80° C. At these temperatures, polar stratospheric clouds containing a mixture of ice, nitric acid and sulphuric acid together with the catalyst chemicals are formed. A reservoir of catalyst chemicals is established in the polar stratospheric cloud layer. The effect is pronounced over the South Pole, where the circulation is effectively concentric around the pole and temperatures can reach the -80° C values. As the earth proceeds in its orbit and polar regions are turned toward the sun, the reservoir warms and the depleting chemicals are released. Destruction of ozone proceeds rapidly within and adjacent to the reservoir region. The circulation pattern also weakens around the reservoir, allowing ozone-depleted air to mix with ozone-enriched air that is moving poleward from the equator. In the northern hemisphere, with its greater land mass, the circulation pattern has a much greater degree of irregularity than in the southern hemisphere, and temperatures are not as extreme as those over the South Pole. For this reason, there is less ozone destruction in the north polar region than in the south.

The rapid rate of thinning of the ozone layer was first noticed in the south polar region in the mid-1970s. The reduced ozone concentration has become known as the ozone hole, though it is not actually a hole; it is a zone where ozone concentrations are up to 40 percent lower than normal values. These lower concentrations allow much more of the damaging ultraviolet radiation to reach the earth's surface. Because of the increased health risks of solar ultraviolet radiation, governments agreed to take steps to deal with the ozone depletion issue. In 1979, many countries, including Canada, banned the use of CFCs as aerosol propellants. In 1987, an agreement known as the Montreal Protocol on Substances that Deplete the Ozone Layer was signed. The agreement required participating countries to reduce their production of CFCs. The agreement was later strengthened to phase out all CFC production.

The Changing Atmosphere

Actions taken under the Montreal protocol are having a positive effect. Measurements in the high atmosphere now show that the concentrations of some of the ozone-depleting substances are slowly starting to decrease. There is, however, a long way to go. It will take until the middle of the 21st century for chlorine levels to return to the values measured in the 1970s when the south polar ozone hole first started appearing. Current projections by Canadian scientists indicate that global total ozone levels will return to 1960s averages by about 2060. The recovery assumes that countries will continue to adhere to the terms of the Montreal protocol and that no other long-lived substances will be found to cause ozone depletion. The possibility of increased NO_X emissions is still a concern because consideration is being given to allowing more high-flying supersonic transport aircraft. Another concern is the potential effect that global warming may have on stratospheric ozone in the polar regions. This is an ongoing topic of Canadian research.

Stratospheric Ozone and Ontario

In Ontario, the thinning ozone layer will continue to be a concern. In late winter to early summer, when ozone thinning is most noticeable over the province, the public is vulnerable to increased levels of ultraviolet radiation. At that time of year, the need to take extra protective measures is especially important for people who enjoy spring skiing. The intensity of UV radiation at alpine levels is enhanced by the thinner atmosphere and by the additional solar reflection off snow surfaces.

Climate Change

Ontarians are well aware of the changes in weather from day to day. We know what to expect from our climate from year to year, but we also know that there will be some degree of variability. Some winters have more snow, some summers are hotter and some have more rain than others. Some years seem to have more severe summer storms, and some years are quite placid. This wide range of variability is the norm. When averages of temperature, rainfall or any of the other atmospheric parameters are calculated at an observing station or are averaged over a region, much of the variability disappears. However, what is apparent from these averages is that there is an undeniable upward trend in temperature with time.

People are finding ways to protect themselves from harmful sun overexposure.

The Changing Atmosphere

Ice Age Temperatures—Greenland and Antarctica

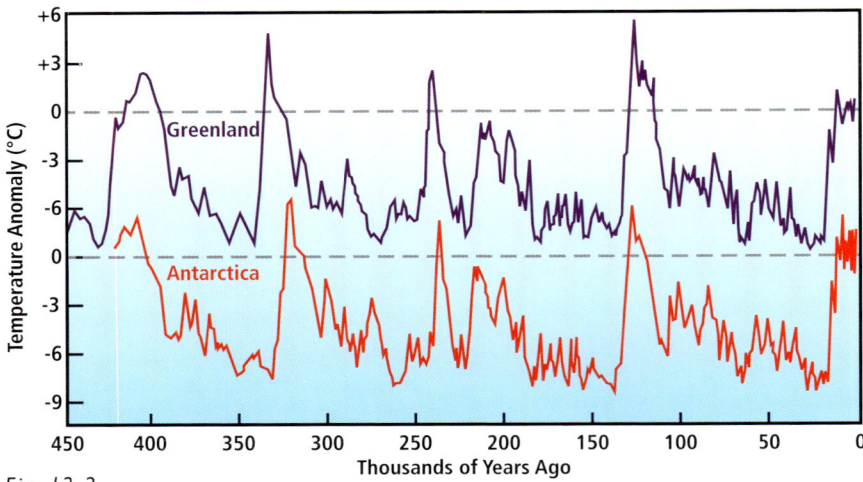

Fig. 12-2

Is the climatic average moving to some new state, and if so, why? What will this new state mean for our lifestyle and for the environment? Can we change the trend? What measures will society need to take to deal with the change? What will be the cost of these measures? These are just some of the many questions that are raised by the climate change issue.

Temperature is the most useful parameter to use in studies of climate and its changes. Temperatures are reasonably consistent over an area, and the temperature record can usually be readily established. The instrumented record goes back to 1840 in Ontario. It is only marginally longer at other sites around the world. Prior to the instrumented record, what can we say about the temperature and climatic conditions? Scientists have developed a number of techniques to deduce past conditions.

We have anecdotal information about storms, cropping practices and settlement patterns that allow us to infer what temperature conditions were like over the last few centuries. From the examination of tree ring widths, we have reasonably direct evidence of the length and productivity of growing seasons as far back as several hundred years. The analysis of lake and ocean bed deposits provides a wealth of information about plant forms that were present on adjacent lands or of plants and animals that inhabited the water bodies. This evidence from around the world helps push the climatic record back several thousand years. Carbon dating is a sophisticated tool that is used to look back about 40,000 years, and it is accurate to within about 100 years. The characteristics of ice in glaciers can also be used to infer what temperatures were like at the time the snow fell on glacier surfaces. Ice cores extracted from Antarctica and

The Changing Atmosphere

from Greenland glaciers have been used to determine what temperatures have been like over the past 600,000 years.

The climate has undergone many natural transitions in the past. Ontario was once covered with lush tropical growth and inland oceans. In sharp contrast, ice sheets have spread across the North American continent with a fair degree of regularity, every 100,000 years or so.

There are several theories that describe why past climatic changes occurred. The first theory relates to the all-important link between the earth and the sun. The earth travels around the sun in a slightly elliptical orbit, and the earth's axis of rotation is tilted at an angle of 23 degrees. Serbian mathematician Milutin Milankovitch examined the regular changes in the degree of orbit elipticity, rates of change in the axis tilt and changes in the time of year that the earth is closest to the sun. With this information, he developed a schedule that could explain the onset and termination of past ice ages.

Shorter-term fluctuations in solar energy output are also well known. The sun goes through a cycle in which sunspots come and go every 11 years. When the number of sunspots is at a maximum, there is a slight elevation in the rate of solar energy output. The increase in radiation is only slightly above the normal levels, though, amounting to less than 0.2 percent over the 11-year sunspot cycle. The sun's energy output may have deviated for periods of a half-century or so when there was an almost total absence of sunspots. The associated solar output decrease has been linked with a cold period in Europe in the 16th and 17th centuries, known as the Little Ice Age. L'Anse aux Meadows, at the northern tip of Newfoundland, was occupied almost 500 years before Columbus arrived in the "New World" in 1492. The site was inhabited for less than 10 years, beginning with Leif Erikkson's arrival around the year 1000 AD. Records indicated that the weather turned colder and the Vikings abandoned their foothold in North America.

213

The Changing Atmosphere

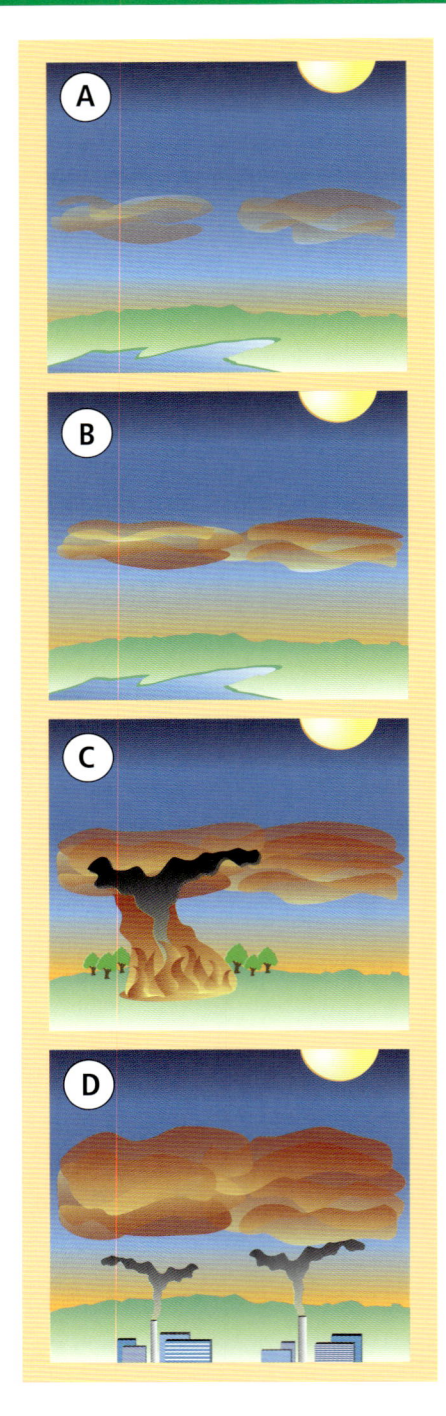

Earth's Energy Balance and the Greenhouse Effect

The temperature of the atmosphere is linked to the balance between incoming solar radiation and the properties of the earth's surface and the surrounding atmospheric envelope. Structures, water and life forms on the earth's surface absorb a large fraction of the total incoming solar radiation. Objects that absorb radiation are warmed, and they in turn radiate energy in the infrared wavelength. Much of this infrared radiation is directly absorbed by radiatively active gases, which then warm and in turn radiate energy. All this energy transfer, absorption and retransmission are in a general state of balance over the earth. This balance is maintained by the meridional circulation. The global energy balance

The Greenhouse Effect

Fig. 12-3
A - sunlight travels though the atmosphere and strikes the earth's surface, which warms and radiates infared energy.
B - most of the infared heat is absorbed by the so-called greenhouse gases in the atmosphere. The gases warm and then radiate heat energy back to the earth.
C - burning forests and other fires release greenhouse gases, especially carbon dioxide, into the atmosphere. This is the natural greenhouse effect, which has given the earth an average surface temperature near 16° C.
D - additional carbon dioxide and other greenhouse gases are released to the atmosphere from fossil fuel combustion in factories and power plants and from transportation. This anthropogenic greenhouse effect is warming the earth and atmosphere at unprecedented rates.

results in an annual average temperature of about 16° C over the entire planet. Of key importance is the role of the radiatively active gases. In 1824, famous French physicist and mathematician Joseph Fourier first proposed the concept that gases in the atmosphere trap heat from the sun. This process has been termed the greenhouse effect, though that is something of a misnomer—greenhouses prevent the mechanical movement of heated air, whereas atmospheric gases reduce the radiative loss of heat. It is estimated that without greenhouse gases, the average global temperature would be about -18° C, which is too cold to support life as we know it. Mars has a thin atmosphere with a low concentration of water vapour and other greenhouse gases; temperatures on the planet are below -50° C. Venus, on the other hand, has a dense atmosphere with high proportions of radiatively active gases. The greenhouse effect on Venus maintains temperatures above 400° C.

Many gases play a role in the natural greenhouse effect on earth, and the most important is water vapour. Water molecules are continuously cycling between the atmosphere and water surfaces, changing phase from vapour to liquid to solid. The evaporation, sublimation and condensation processes all involve substantial energy exchanges, which drive much of the weather we experience. The average water content of the atmosphere changes little and is kept in balance through the evaporation-precipitation cycle. However, how much water the atmosphere can hold is dependent on temperature—a warmer atmosphere tends to hold more water vapour, providing a positive feedback mechanism to heat the atmosphere further.

Other important greenhouse gases are carbon dioxide (CO_2), methane (CH_4) and nitrous oxide (N_2O). These gases have a natural origin, but they are also produced through human activity. Carbon dioxide is released from combustion processes such as forest fires and through fossil fuel combustion in heating systems, engines and boilers. Carbon dioxide is also produced in respiration and through the aerobic decay of vegetation. Methane is produced through anaerobic decay of vegetation and in the digestive tract of ruminants. It is the primary constituent of natural gas, is contained in coal beds and is a minor by-product of the combustion of carbon-based fuels. Nitrous oxide is primarily a product of combustion, though it is also released to the atmosphere when nitrogen-based fertilizers break down. All these gases have atmospheric lifespans that are much longer than that of water vapour, and this is what makes them of particular concern as greenhouse gases because they tend to accumulate in the atmosphere.

There are other gases that have no natural form in nature but are only produced by humans. Substances such as CFCs, halons and several others are in this category and are potent greenhouse gases, in part because of their long atmospheric lifespans.

The Trend in Global Greenhouse Gases

Atmospheric CO_2 concentrations have been measured for less than a century. Measurements taken at Mauna Loa, Hawaii, since 1956 show a steady upward trend, with mean annual atmospheric concentrations increasing from about 300 parts per million (ppm) to the present value of 380 ppm. Similar trends are noted at Canadian sites and at other locations around the world. Analysis of the atmospheric gases trapped in glaciers and ice sheets provides a longer record. Ice cores from the south polar ice sheet have made it possible to reconstruct the atmosphere's CO_2 concentration extending back more than 600,000 years. Data from these cores shows that CO_2 concentrations were about 270 ppm prior to industrialization. During periods of major global glaciations, global CO_2 concentration levels were about 200 ppm. During the warm interglacial periods, including the present one, CO_2 levels attained peak values of 270 to 300 ppm. Methane and nitrous oxide concentrations show similar long-term trends. In summary, the concentrations of all the key greenhouse gases are far higher now than at any time during the past 600,000 years.

Concerns about Warming

In 1896, while investigating the causes of periodic ice ages, Swedish chemist Svante Arrhenius quantified the relationship between the average concentrations of greenhouse gases and average global temperatures. He concluded that a doubling of atmospheric carbon dioxide concentrations would likely result in an average global warming of 5° to 6° C. Arrhenius realized that the carbon dioxide introduced by industrial emissions was in fact accumulating in the atmosphere, but he concluded that, given the emission rates at the turn of the 20th century, a doubling of CO_2 concentrations would be unlikely for 3000 years. In reality, emissions from new industrial emissions sources are being produced at much greater rates than he envisioned, and a doubling might be expected to occur within 100 years.

There is clear evidence that the earth has undergone a steady warming over the past century. The scientific community has been working to confirm the climate trend and to determine the reasons. Careful examination and re-examination of temperature data collected around the globe reinforces that a warming is underway. The temperature record shows that since the end of the 19th century, the average global temperature has risen about 0.6° C. This warming has not been steady. The instrument record shows that a period of warming from about 1910 to 1940 was followed by a short period of cooling until the mid-1970s. The warming since then has been steady. The variability of the trend has caused much controversy. Some brief cooling periods can be explained by volcanic eruptions such as Mount Pinatubo in 1991, but not all the variability has been explained. Nevertheless, the general trend is decidedly upward, and most scientists agree that this is because of excessive greenhouse gas emissions.

Data has been compiled for the northern hemisphere, consisting of an amalgam of measured temperature records and temperatures inferred from tree ring growth, coral growth and ice core analyses. The period from about 1000 AD to 1900 AD showed a slight cooling. Over the past century, however, the warming has been remarkable.

> **Love comforteth like sunshine after rain.**
>
> —Shakespeare,
> *Venus and Adonis*

The Changing Atmosphere

Atmospheric Carbon Dioxide Trends

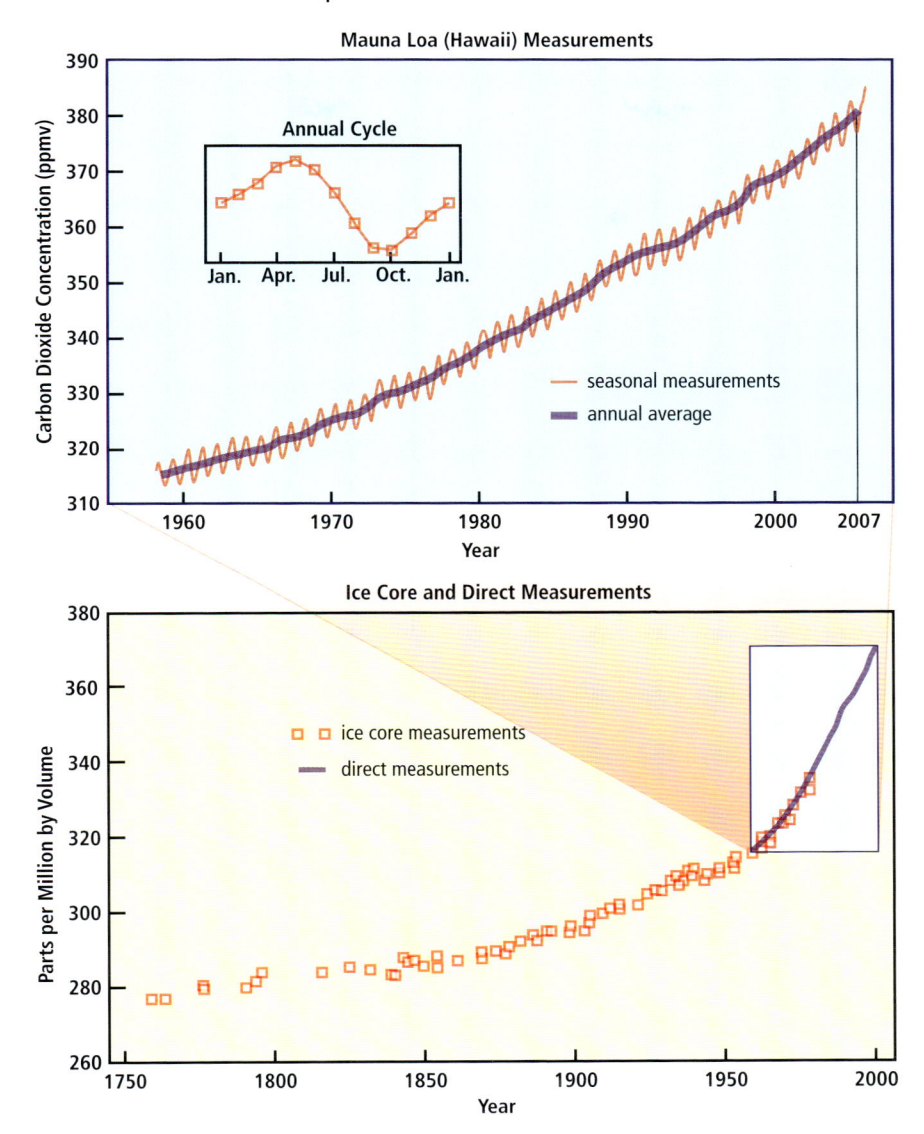

Fig. 12-4 Atmospheric carbon dioxide concentrations as determined from ice cores show a steady but modest increase from 1750 to 1950. Since 1960, instrument measurements and ice core data show the rate of increase has accelerated. Global values are now over 380 ppm.

The Changing Atmosphere

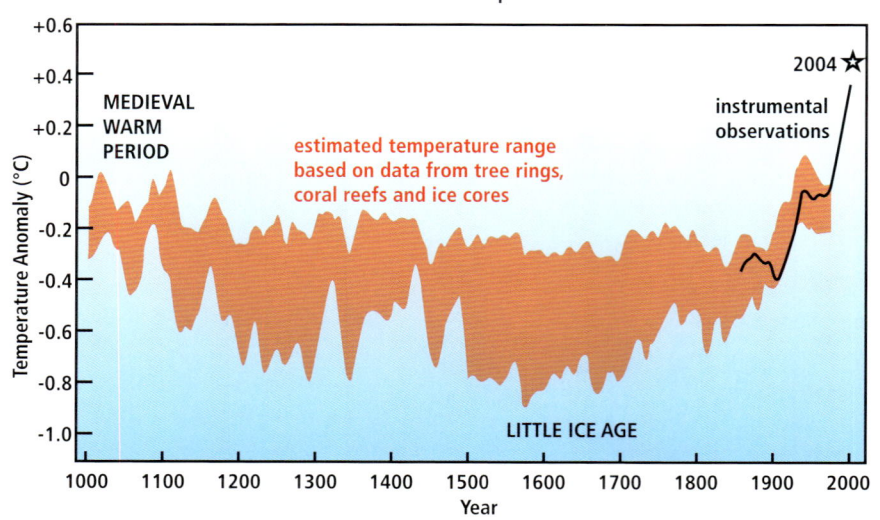

Reconstructed Temperatures

Fig. 12-5 The shaded region shows the range of uncertainty in temperatures inferred from tree rings, coral reefs and ice cores. The solid line is the instrumented record. All data reflect the global averages from 1961 to 1990.

Global Temperatures

Fig. 12-6 The squares are average temperatures computed from thousands of surface stations around the globe. The data reflect averages for the period from 1961 to 1990. The smooth curve is the five-year running mean of the plotted data.

The Changing Atmosphere

Climate Change in Ontario

In Ontario, the temperature record extends back to the late 19th century. When annual temperatures are plotted, there is considerable variation from year to year, but there is evidence of a warming trend. The record for sites in Ontario shows elements of the same trends noticed elsewhere in the northern hemisphere.

Since 1900, the frost-free season has lengthened by 20 to 30 days at many locations across Canada, but the growing season has not changed significantly.

The extreme temperature range has decreased by 4° to 6° C at more than half the stations because minimum temperatures have risen but the warm extremes have not changed much.

The below graphs show the temperature anomalies from 1900–2001 for 210 climate stations across Canada. The red line is the 11-year running mean for the given temperature index. The number of cold nights is certainly decreasing, while the number of warm days shows a more subtle and gradual increase.

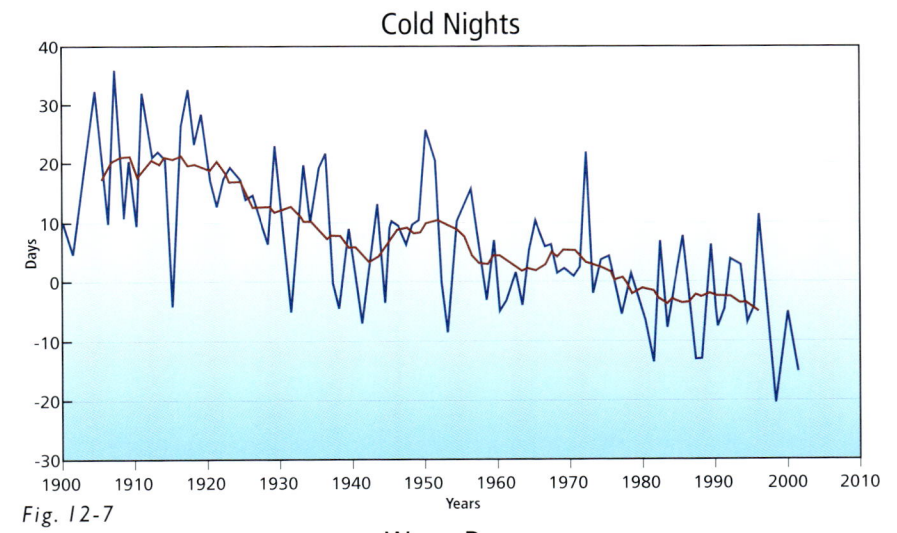

Fig. 12-7

Fig. 12-8

The Changing Atmosphere

The same research study also shows that the annual number of days with precipitation has increased almost everywhere across Canada by 20 to 40 days. The number of days with snow has decreased south of 60° N but has increased substantially in the north and northeast. Weather systems and the storm track are migrating toward the north. The intensity of precipitation events has decreased at most stations, and the ratio of snowfall to total precipitation has certainly decreased in the south.

In the below graphs, precipitation anomalies from 1900–2001 are shown for 210 climate stations across Canada. The green line is the 11-year running mean for the given precipitation index. The number of days with rain is certainly increasing, but there is no trend in the highest 5-day precipitation amount.

Research conducted throughout Ontario also reveals generally more precipitation from 1971–2000 than from 1951–1980. The enhanced precipitation because of snowsqualls in onshore areas is consistent with warmer and longer ice-free periods over the Great Lakes, which could also contribute to greater evaporation from their surfaces.

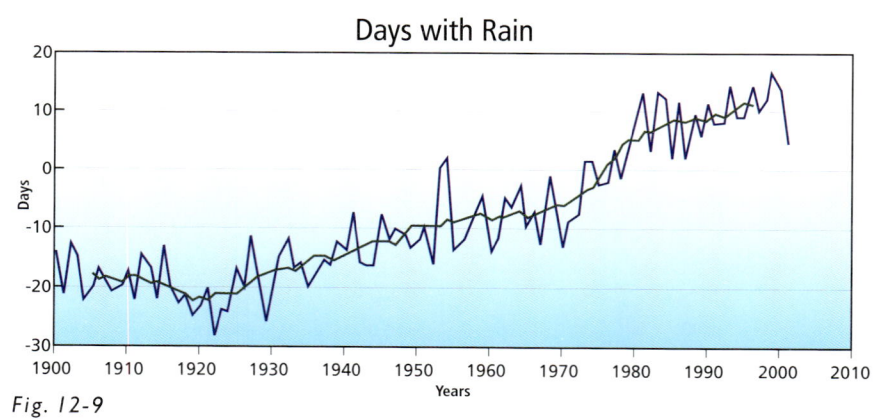

Fig. 12-9

Fig. 12-10

Future Climate

Predicting future climate is a challenging task. The climate system is composed of several components, including the atmosphere, the hydrosphere (oceans, lakes, wetlands and rivers), the cryosphere (glaciers, floating ice) and the biosphere (including all living things). These components of the climate system interact and influence each other. Components of the climate system that have a physical science basis can be described by mathematical relationships. A representation of the climate system can then be made by means of a complex mathematical model formulated to operate on computers. This is known as a climate simulation model. These equations depend on time, so the model can be used to examine how the climate system changes with time. Scientists use this model to develop scenarios of the future climate.

Climate simulation models work much like the weather prediction models described in an earlier chapter, but there are several important differences. Weather prediction and climate simulation models both use the laws of fluid physics to describe the atmosphere and its changes with time. Climate models, however, take into account many additional processes that are not important for short-range weather forecasting. Weather prediction models start from an initial state of the atmosphere and run forward a few days into the future. Climate models run forward for decades, even out to 100 years or more. The climate models run on a coarser scale than weather prediction models—usually with a horizontal grid spacing of hundreds rather than tens of kilometres—and therefore the climate detail is relatively coarse. All the complexities of the linkages between the atmosphere, the hydrosphere and the cryosphere must be calculated using relationships appropriate to the grid scale of the model. The number of calculations necessary to undertake a climate simulation can only be accomplished in a reasonable time frame by using the most powerful computers available. Only a few centres in the world have sufficient computing power and scientific expertise to undertake numerical climate simulations. Some of the more prominent centres include the Hadley Centre of the British Meteorological Office, the Geophysical Fluid Dynamics Laboratory (GFDL) of the National Oceanographic and Atmospheric Administration of the United States, the National Center for Atmospheric Research (NCAR) located in Colorado, and the Earth Simulator Center in Japan. In Canada, Environment Canada conducts climate simulation modelling with a group located at the Canadian Centre for Climate Modelling and Analysis in Victoria, BC.

The output of the climate simulation models can be displayed either in pattern form (i.e., maps) or by displaying the value of some climatic parameter that is computed by the model. Average surface temperatures over the globe, over a hemisphere or even over a region are often displayed. For meteorologists to have confidence that the models are correct, the models must accurately depict our present climate. A good test of their validity is to see how they perform in simulating past climate trends.

The Changing Atmosphere

Model Simulations of Past Climate Trends

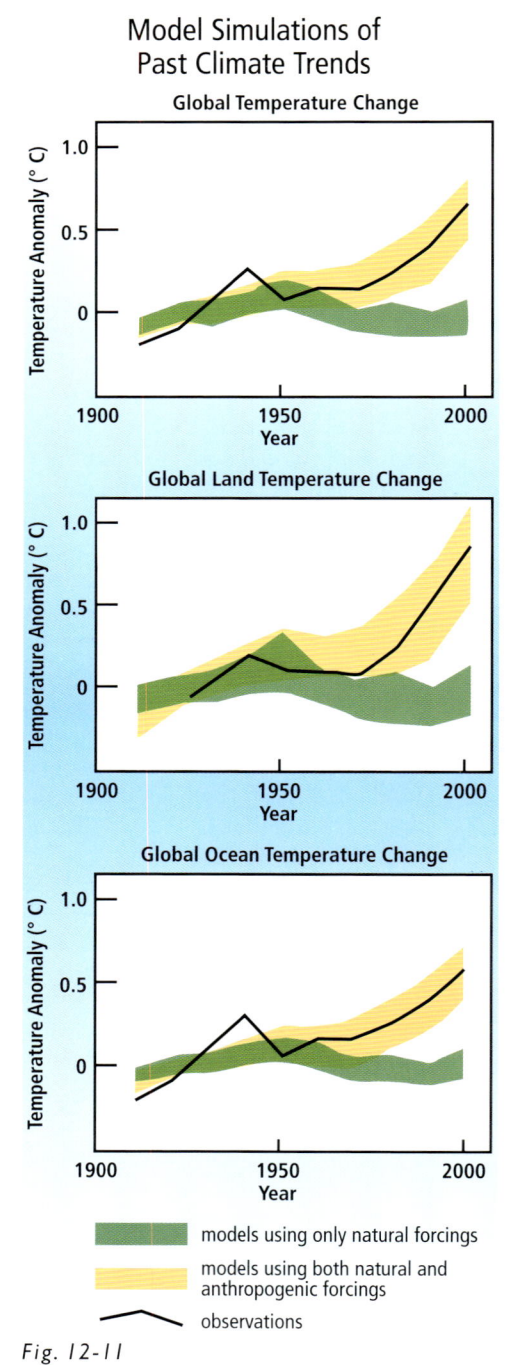

Global Temperature Change
Global Land Temperature Change
Global Ocean Temperature Change

- models using only natural forcings
- models using both natural and anthropogenic forcings
- observations

Fig. 12-11

Figure 12-12 shows how 14 models have performed in simulations of worldwide surface temperature over the oceans and land services throughout the past century. Also shown is the average of observations calculated in the same areas over the same period. The ranges of the model simulations, shown as shaded bands, depend on the assumptions used in the formulation of the model physics. The graph shows that models best replicate the observed global temperature trend when increases in anthropogenic greenhouse gas emissions are incorporated in model simulations.

In 1988, the United Nations established the Intergovernmental Panel on Climate Change (IPCC) to investigate whether human activity is causing climate change. The emphasis over the past three decades has been on the role of greenhouse gases in changing the short-term climate. Climatic simulation models are used to undertake experiments using a range of greenhouse gas emission scenarios. Figure 12-13 shows the results of several emission scenario experiments undertaken using a number of climatic simulation models. The models show that there is a wide range of possible outcomes for the future climate, depending on the emissions scenario. The models project that by the end of the 21st century, the earth will have warmed between 1.1° C and 6.4° C relative to the global temperature at the end of the 20th century.

The Changing Atmosphere

Multi-modal Averages and Assessed Ranges for Surface Warming

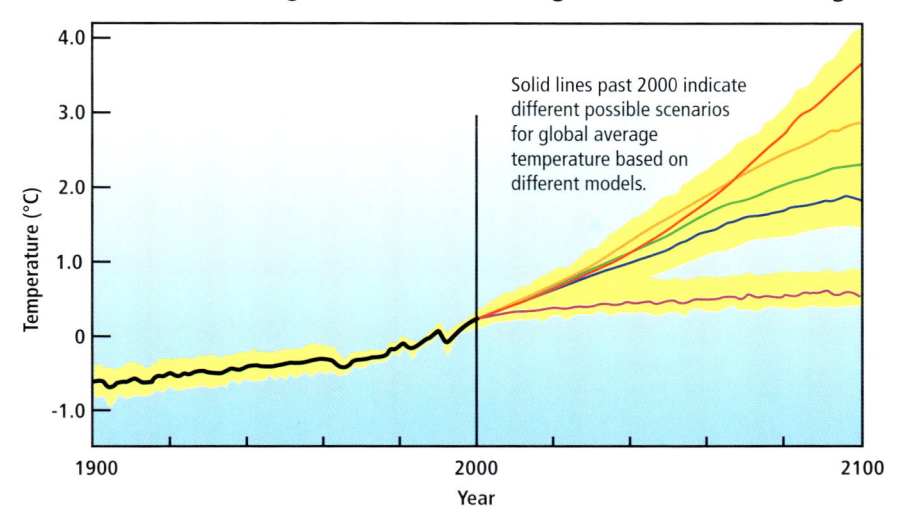

Fig. 12-12 The yellow shaded band shows the results of simulations of global average surface temperature from many different climate simulation models. The solid curve prior to 2000 is the observed temperature. Each solid line past 2000 is the average of model simulations for one emission scenario.

Ontario's Future Climate

The present trends in climatic observations together with climate model simulations suggest that the world's climate will undergo further warming during this century. A few general observations can be made about weather trends in Ontario based on climate projections from the Canadian climate simulation model. The province will likely undergo a continued gradual warming throughout this century with a greater warming rate in the north than in the south. One can anticipate more frequent warm spells in summer, and nighttime temperatures will warm more than the daytime maximum temperatures—about twice as fast, at 0.2° C per decade. In winter, a trend toward fewer cold spells is expected. This does not mean that we will not experience any more cold days or even cold winters. Occasional cold winters will still happen, but they can be expected to occur less frequently than we have seen in the past. Generally, the snows of winter will arrive later, and the warmth of spring will arrive earlier.

The climate model simulations suggest slight increases in average precipitation amounts over Canada as well as a trend to more frequent and extreme precipitation events. The combination of slight increases in annual precipitation together with significant warming suggests a greater frequency of extreme summer heat waves. With the increased temperatures and associated evaporation, there is also a trend toward a more arid climate with a higher probability of drought conditions.

The Changing Atmosphere

Decades of pollution such as this have added to the climate change problem.

Because climate models typically use a fairly coarse grid, simulating climate changes at a regional scale is a major challenge. Regional climate models are being developed to deal with this issue and are being tested throughout Canada. The outputs from the Canadian model are eagerly anticipated. More definitive studies will undoubtedly follow, providing better resolution of climate trends and impacts. This will then allow for an improved understanding of the options available and the costs associated with accommodating a changing climate.

The Changing Atmosphere

The Global Future Climate

Some speculative studies have been undertaken to look at the global impacts of climate changes. Briefly, they suggest the following:

- the world's boreal forests face an increased fire risk because of invasive pests, the drying climate and increasing thunderstorms

- water needs will outstrip supply. The loss of water could be severe because of changes in evaporation and precipitation patterns

- low-lying countries will experience coastal flooding as sea levels rise, and coastal land will be lost

- tropical diseases such as malaria will move northward, where populations have little or no immunity.

The weather and climate change affect us all. The economic and human losses from severe weather can be tragic. As the population of Canada grows, vulnerable infrastructure increases along with chances that severe weather events will be even more catastrophic.

> *The Great Lakes system is the largest source of fresh surface water on earth. Only 1% of the Great Lakes' volume is renewed on an annual basis; using a higher volume per year will tax the lakes, reducing their water levels beyond what can be replenished naturally.*

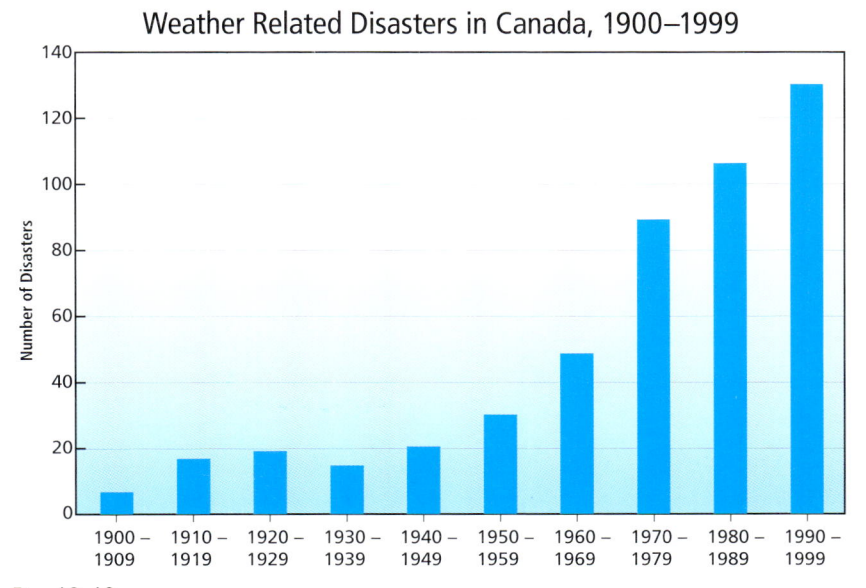

Weather Related Disasters in Canada, 1900–1999

Decade	Number of Disasters
1900–1909	~6
1910–1919	~17
1920–1929	~19
1930–1939	~15
1940–1949	~21
1950–1959	~30
1960–1969	~49
1970–1979	~89
1980–1989	~107
1990–1999	~130

Fig. 12-13

The Changing Atmosphere

Impact of Climate Change

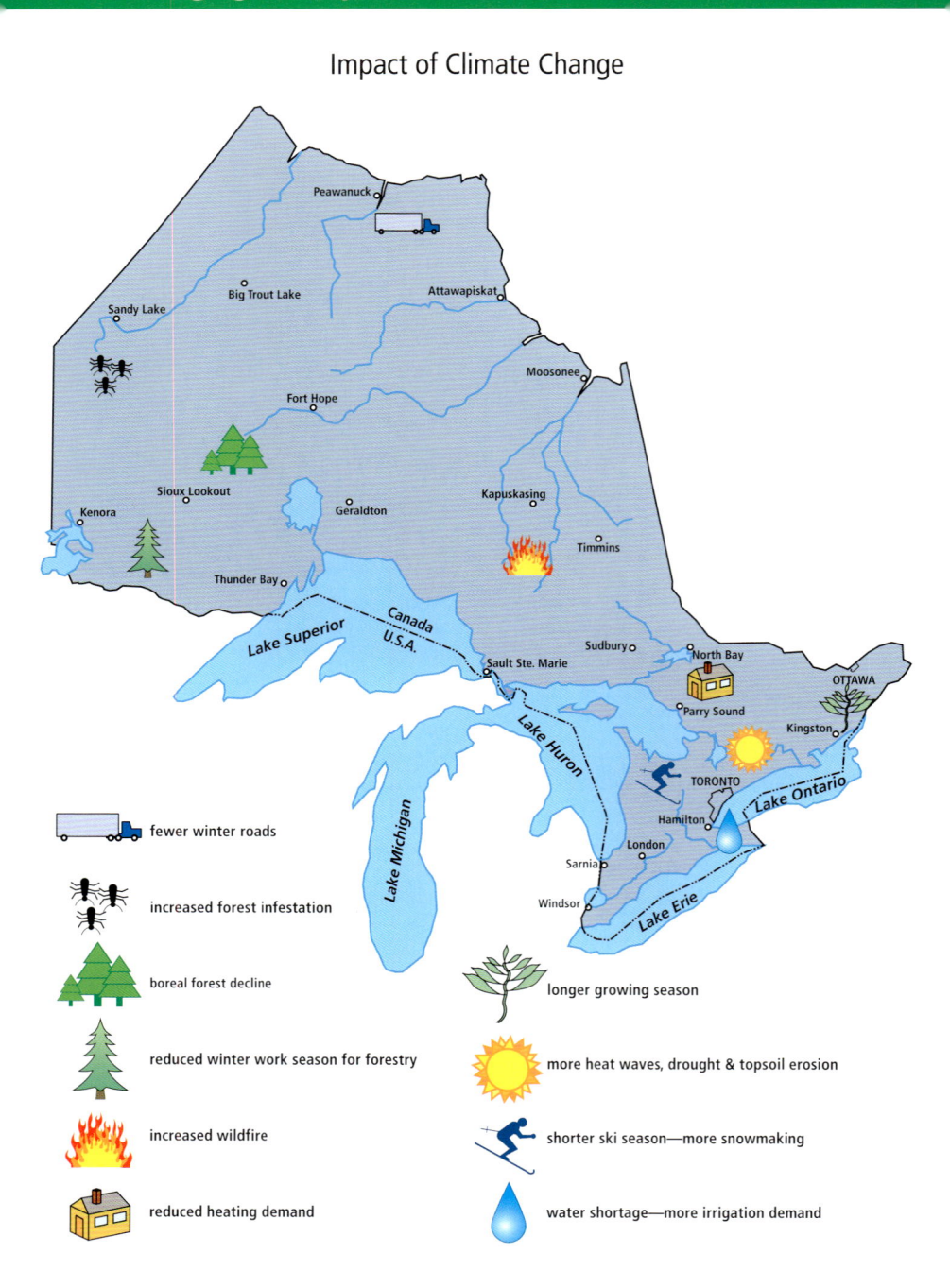

Legend:
- fewer winter roads
- increased forest infestation
- boreal forest decline
- reduced winter work season for forestry
- increased wildfire
- reduced heating demand
- longer growing season
- more heat waves, drought & topsoil erosion
- shorter ski season—more snowmaking
- water shortage—more irrigation demand

Fig. 12-14

The Changing Atmosphere

A Plan for All Seasons

In *Weather of Ontario* we've provided the basics for you, the reader, to understand, anticipate and appreciate the many facets of the weather of Ontario. Such knowledge can help you to plan for each unique and special season and can empower you to take action appropriate to the weather. A perfect forecast is wasted if not acted upon.

Weather is a force of nature largely beyond our control. But with understanding and prediction, we can appreciate weather and adapt our plans for safety and security. Severe weather may be especially exciting but it is not something to experience first hand. You may want to see a tornado before you die, but not *just* before you die.

Meteorology is a wonderful science and a most satisfying career. The science continues to advance and our understanding of the atmosphere improves with new observation systems and ongoing research. The vast potential to serve society and help heal the atmosphere is the reward. Plus, meteorology is fun and there are tons of forecaster jokes like "50% puts you at the head of your class."

The weather and seasons are full of promise, inspiration and hope. Mankind has certainly altered the environment. But mankind can also take action to undo the wrongs of the past. Everyone acting individually can collectively make an impact. As one small example, on my 25-acre paradise at the crest of the Oak Ridges Moraine, I have built and maintained what is probably the highest density of Peterson bluebird houses on the planet. As a result, we now enjoy the company of countless numbers of these once rare birds. "Build it and they will come." Other species have returned along with the bluebirds. There are now 3000 trees where there were only seven. Honeybees forage the wild flowers, which have been chemical free since we "bought the farm." With proper stewardship of the environment, nature continues to heal itself. Everyone can have the same green impact regardless of where they live.

> The natural world is spectacular. It is to be embraced and appreciated but protected at the same time. And there is weather everyday! Enjoy it.
>
> —Phil Chadwick

Glossary

accretion: the growth of a hydrometeor through collision with supercooled cloud drops.

air mass: an extensive body of air within which temperature and moisture at a constant height are essentially uniform.

albedo: the rate of reflectivity of incoming solar radiation at the earth's surface.

ambient air quality: air pollutant concentration levels for outdoor air.

anticyclone: an atmospheric pressure system that has relatively high pressure at its centre and, in the northern hemisphere, has winds blowing around the centre in a clockwise direction; also called, simply, a high.

azimuth angle: the angle in degrees from north measured in a clockwise direction.

biogeoclimatic zone: a geographical area with a relatively uniform climate and associated vegetation.

blue jet: a weakly luminous blue-coloured discharge that moves upward from the top of a thunderstorm.

boreal: relating to cool northern climates and the associated plant and animal life.

cirrus: a principal cloud type in the form of thin or feather-like clouds in the upper atmosphere, made up of ice crystals.

convection: the process generally associated with warm, rising air and the formation of cloud in an unstable atmosphere. Convection cells can form in any fluid, including the Earth's atmosphere.

Coriolis effect: the apparent deflection of moving objects as a result of the earth's rotation. Objects deflect to the right in the northern hemisphere and to the left in the south.

cumulus: a cloud in the form of detached domes or towers that appear dense and have well-defined edges; has a flat, nearly horizontal base with a bulging upper portion that resembles cauliflower.

cup anemometer: a wind-measuring instrument consisting of three or four hemispherical or conical-shaped cups mounted on horizontal arms that rotate around a vertical axis.

cyclone: an atmospheric pressure system that has relatively low pressure at its centre and, in the northern hemisphere, has winds blowing around the centre in a counter-clockwise direction; also called, simply, a low.

discretization: the method of representing a continuously varying parameter with a discrete set of values.

downburst: localized area of damaging winds caused by air rapidly descending from a thunderstorm.

fetch: the length of water over which the wind has blown.

freezing level: the lowest altitude in the atmosphere where the air temperature is 0° C.

front: the boundary region along the earth's surface that separates air masses; frontal types are characterized by the nature of the cold air mass, the direction of the movement of the front and its stage of development.

frontal lift: the upward displacement of an air mass of lower density by an air mass with a higher density.

Glossary

frontal surface: the interface above the earth's surface that separates different air masses.

fulgurite: a glassy tube formed when a lightning stroke terminates in dry, sandy soil and fuses the sand.

funnel cloud: a condensation cloud, typically funnel-shaped and extending outward from a cumuliform cloud, associated with a rotating column of air (a vortex) that may or may not be in contact with the ground. If the rotation is violent and in contact with the ground, the vortex is a tornado. Funnel clouds can occur through a variety of processes in association with convection. For example, small funnel clouds can be seen extending from small, dissipating cumulus clouds in environments with significant wind shear.

geostrophic wind: the air motion that results from air pressure differences.

gradient: the change of a meteorological parameter per unit of distance.

graupel: a form of ice created in the atmosphere when supercooled water droplets coat an ice crystal.

gustnado: colloquial expression for a short-lived, shallow, generally weak tornado found along a gust front. Gustnadoes are usually visualized by rotating dust.

hydrometeor: a general term for any type of water or ice particle.

isobar: a line on a chart or diagram drawn through points that have the same barometric pressure.

jet stream: a zone of relatively strong winds concentrated within a narrow band in the atmosphere.

lapse rate: the rate of decrease of atmospheric temperature with height.

mares' tail: a cirrus cloud that has long, wispy strands extending from a tufted end.

mesocyclone: an updraft column of air with a diameter of 2 to 10 kilometres that rotates around a vertical axis in a large thunderstorm; the rotating column can lead to the formation of tornadoes.

mesopause: the atmospheric layer located at a height of around 90 kilometres, at the top of the mesosphere, where the coldest atmospheric temperatures occur.

mesosphere: the atmospheric layer above the stratosphere, characterized by temperatures that decrease from the base of the layer to the top.

N-region: the negatively charged region in the lower portion of a thunderstorm cloud.

nonsupercell tornado: a tornado that occurs with a parent cloud in its growth stage and with its rotation originating in the boundary layer. The parent cloud does not contain a pre-existing mid-level mesocyclone. Landspouts (dust devils), elevated funnels and waterspouts not associated with a supercell are examples of nonsupercell tornadoes.

occluded front: formed when a cold front overtakes a warm front during the latter stages of frontal development.

planetary (long) waves: a wave-shaped pattern in the westerly wind belt characterized by long length; four or five waves are typically found around the hemisphere, and the wave pattern generally moves slowly toward the east with a speed much slower than the speed of the westerly winds.

Glossary

planetary boundary layer: the bottom layer of the troposphere in contact with the earth's surface.

p-region: the positively charged region near the bottom of a thunderstorm cloud.

P-region: the positively charged region near the top of a thunderstorm cloud.

pressure gradient force: the force acting on air caused by differences in air pressure; the strength of the force is proportional to the difference in air pressure.

psychrometer: an instrument that measures humidity using two thermometers, one of which has its sensing bulb covered by a water-saturated muslin jacket.

retrogression: westward movement of weather systems at middle latitudes, instead of the usual eastward movement.

red sprite: a large-scale luminous flash that appears directly above an active thunderstorm and extends from a cloud top to heights of about 90 kilometres; it is delayed a short time after a cloud-to-ground lightning stroke.

ridge: an elongated area of high atmospheric pressure on a weather chart.

rime: milky-coloured ice formed when super-cooled water drops freeze on contact with a surface and create a mass of tiny balls with air spaces between them.

shortwave trough: small-scale disturbances embedded in the upper air flow.

Introduction

silver thaw: colloquial expression for a deposit of glaze on trees and other objects from a fall of freezing rain.

smog: originally defined as a mixture of smoke and natural fog, but now describes a mixture of air pollutants that are emitted by automobiles and industrial sources and are acted upon by the sun.

snow pillow: an instrument used to estimate snow pack by measuring the pressure exerted by a mass of overlying snow.

solar wind: a stream of charged particles flowing outward from the sun.

stepped leader: the column of highly ionized air that intermittently advances downward from a thunderstorm cloud and establishes a channel for a lightning stroke.

stratopause: the top of the stratosphere, characterized by a reversal of the rate of change of temperature with height.

stratosphere: the atmospheric layer from about 10 to 50 kilometers above the earth, characterized by temperatures that warm from the base to the top of the layer.

stratus: a cloud layer with a fairly uniform base.

streamer: an upward-advancing column of highly ionized air that moves from a point on the earth's surface toward a stepped leader.

supercell: a large, severe thunderstorm with rotating updraft that usually lasts several hours and produces heavy rain and hail, strong winds and occasionally spawns tornadoes.

supercell tornado: a tornado that occurs within a supercell that contains a well-established mid-level mesocyclone.

synoptic scale storm: a large-scale atmospheric system or weather event that has a horizontal dimension of greater than 300 kilometres.

thermistor: a temperature sensor in which the electrical resistance varies in proportion to the temperature; the name is a combination of "thermal" and "resistor."

thermosphere: the upper atmosphere above 100 kilometres characterized by low gas density and temperatures increasing with height.

tornado: a violently rotating column of air, in contact with the ground, either pendant from or under a cumuliform cloud.

tripole distribution: the characteristic distribution of positive and negative charge centres in thunderstorm clouds.

tropopause: the transition zone in the upper atmosphere between the troposphere and the stratosphere, characterized by an abrupt reversal of the thermal lapse rate.

troposphere: the atmospheric layer from the earth's surface to about 8 to 12 kilometers, characterized by decreasing temperature with height and substantial water vapour; the region in which most weather occurs.

trough: an elongated area of low atmospheric pressure on a weather chart.

trowal: a trough of warm air aloft.

waterspout: any tornado over a body of water, but in its most common form a nonsupercell tornado over water.

Resources

Selected References

American Meteorological Society. "Glossary of Meteorology". http://amsglossary.allenpress.com/glossary

Barry, R.G. and R.J. Chorley. (1971). *Atmosphere, Weather and Climate.* London: Butler & Tanner Ltd.

Bates, David V. and Robert B. Caton. (2002). *A Citizen's Guide to Air Pollution* (2nd ed.). Vancouver: David Suzuki Foundation.

Cessna Pilot Center. (1984). *Canadian Manual of Flight.* Denver: Jeppesen & Co.

Drew, John. (1855). *Practical Meteorology.* London: John Van Voorst.

Fletcher, N.H. (1962). *The Physics of Rainclouds.* New York: Cambridge University Press.

Flohn, Hermann. (1969). *Climate and Weather.* New York: McGraw-Hill Book Company.

From the Ground Up. (1987). Ottawa: Aviation Publishers Co. Ltd.

Gdezelman, Stanley David. (1980). *The Science and Wonder of the Atmosphere.* New York: John Wiley & Sons.

Hare, F. Kenneth and Morley K. Thomas. (1974). *Climate Canada.* Toronto: John Wiley & Sons Canada Ltd.

Harrison, Louis P. (1942). *Meteorology.* New York: National Aeronautics Council, Inc.

Heidorn, Keith C., PhD. "The Weather Doctor" http://www.islandnet.com/~see/weather/doctor.htm

Holton, James. R, Judith A. Cury and John A. Pyle (Eds.). (2002). *Encyclopedia of Atmospheric Sciences.* San Diego: Academic Press.

Intergovernmental Panel on Climate Change. (2007). "Summary for Policymakers." *Climate Change 2007: The Physical Science Basis* (contribution of Working Group I to the Fourth Assessment Report of the Intergovernmental Panel on Climate Change). Cambridge: Cambridge University Press.

Meteorological Branch. (1964). *Weather Ways.* Ottawa: Department of Transport, Queens Printer and Controller of Stationery.

Meteorology Education and Training, MetEd, Operated by the COMET Program.
http://www.meted.ucar.edu/index.htm

The National Center for Atmospheric Research, NCAR RAP Real-Time Weather Data.
http://www.rap.ucar.edu/weather/

NorLatMet, Northern-Latitude Meteorology, NorLat.
http://www.meted.ucar.edu/norlat.php

Phillips, David. (1990). *The Climates of Canada.* Ottawa: Supply and Services Canada.

Rakov, Vladimir A. and Martin A. Uman. (2003). *Lightning: Physics and Effects.* Cambridge: Cambridge University Press.

Shaefer, Vincent J. and John A. Day. (1981). *A Field Guide to the Atmosphere.* Boston: Houghton Mifflin Co.

Spiegel, Herbert J. and Arnold Gruber. (1983). *From Weather Vanes to Satellites.* New York: John Wiley & Sons.

Stanley, George F.G. (1955). *John Henry Lefroy: In Search of the Magnetic North.* Toronto: MacMillan Company of Canada Ltd.

Uman, Martin A. (1986). *All About Lightning.* New York: Dover Publications, Inc.

Weatheroffice, Environment Canada
http://www.weatheroffice.gc.ca/

Weisberg, Joseph S. (1981). *Meteorology: the Earth and Its Weather.* Boston: Houghton Mifflin Co.

Williams, Jack. (1992). *The Weather Book.* New York: Vintage Books.

Williamson, S.J. (1973). *Fundamentals of Air Pollution.* Reading: Addison Wesley Publishing Co.

Zielinski, Gregory A. and Barry D. Keim. (2003). *New England Weather, New England Climate.* Lebanon: University Press of New England.

Index

A
accretion, 57
acid deposition, 204
acid rain, 204
acidification, 204
acoustic sounder. *See* SODAR
aerological measurements, 173
air glow, 82
air mass, 20, 114
    arctic, 114
    continental arctic, 20, 114
    continental polar, 20
    maritime arctic, 20, 114
    maritime polar, 114
    maritime tropical, 20, 114
Alberta Clipper, 24, 152
ambient air quality objectives, 205
ammonia, 196
anemometer
    cup, 170
    hot-film, 171
    hot-wire, 171
    Patterson three-cup, 170
    propeller-and-vane, 171
    Robinson cup, 170
    sonic, 171
arctic sea smoke, 35
Arrhenius, Svante, 216
atmosphere
    composition of, 11
    vertical variation of, 14
atmospheric pressure, 12, 18
aurora borealis, 66, 80–82
azimuth angle, 180

B
barometer
    aneroid, 169–170
    mercury, 169
Bergeron process. *See* raindrop formation process
blizzard, 148–150
blue jet, 74
Boltzman's law, 177
breeze
    lake, 84–85
    land, 85
bright water horizons, 109
bromine, 209

"Buddha's fingers." *See* crepuscular rays
buffering capacity
    high, 204
    low, 204
buoy
    drifting, 172
    moored, 172

C
carbon dioxide, 195, 215
carbon monoxide, 195
cell, 51, 90
Celsius, Anders, 166
chlorofluorocarbons, 209
circulation
    anticyconic, 16
    atmospheric, 16
    cyclonic, 94
    Hadley, 16
    meridional, 214
cirrus shield, 46
climate record, 212
climate simulation model, 221
cloud droplet, 26, 55
cloud electrification process, 67
    convective process, 68
    graupel-ice process, 69
cloud street, 37
clouds. *See also* cirrus shield, cloud street, contrail, fibrous cirrus veil, fog, Helmholtz waves, mares' tails
    altocumulus, 39
    altocumulus castellanus, 40
    alto form, 30
    altostratus, 38
    cirrocumulus, 44
    cirrostratus, 43
    cirrus, 30, 42
    convective, 63, 68
    cumulonimbus, 50
    cumulus, 30, 47
    cumulus congestus, 48
    cumulus humilis, 47
    cumulus pileus, 49
    high, 30, 42–46
    lenticular, 41
    low, 30, 31–37
    middle, 30, 38–41
    mammatus, 53
    multicell cumulonimbus, 52
    nimbostratus, 36
    nimbus, 30

Index

noctilucent, 45
pulse type cumulonimbus, 51
roll, 94
standing wave, 41
stratocumulus, 32
stratus, 30, 31, 32
stratus fractus, 31
supercell cumulonimbus, 52
towering cumulus, 48
turbulent stratocumulus, 37
wall, 53
wave, 41
cold low, 134
collision theory. *See* raindrop formation process
contrail, 46
convection, 29
convective instability, 93
convective mixing, 15
convective process. *See* cloud electrification process
Coriolis
  effect, 17, 83
  force. *See* Coriolis effect
  Gustave-Gaspardè, 17
corona,
  positive, 68
crepuscular rays, 109
curvature in air flow, 84

D

dew point temperature, 26
dewcel, 168
diamond dust, 105
discharge. *See also* blue jet, red sprite, streamer
  coronal, 79–80. *See also* St. Elmo's fire
  negative, 70
  positive, 74
  single stroke, 72
discretization, 183
Doppler
  principle, 176
  radar, 176
  shift technique, 188
downburst, 91
downdraft, 94
Drive Clean Program, 206
drizzle, 57, 58
  freezing, 58
drought, 138–139
dust devil, 96

E

electrosphere, 66
El Niño, 116–120
El Niño-Southern Oscillation, 118
ENSO. *See* El Niño-Southern Oscillation

F

Fahrenheit, Gabriel Daniel, 166
fetch, 189
fibrous cirrus veil, 50
fog, 33–35. *See also* mist
  advection, 35
  radiational, 34
  steam, 35. *See also* arctic sea smoke
forecast
  marine, 189–190
  short-range, 184
  summer severe weather, 187–188
  Terminal Aerodrome, 186, 192
  thunderstorm, 74
  tornado, 188
  winter severe weather, 188–189
forest fire, 77, 205
Fourier, Joseph, 215
frictional effects, 83
front, 20, 56
  cold, 20
  gust, 90–91
  warm, 20
frontal
  concept, 20
  region, 20
  surface, 27
  zone, 20, 116
frontal lift, 16, 27, 56
  cold, 27
  warm, 27
frost-free period, 129
F-scale. *See* Fujita scale
Fujita scale, 92
Fujita, Theodore, 92
fulgurite, 78
funnel
  cold air, 97
  condensation, 53

G

global warming, 216
global weather prediction grid, 184–185
Graphical Forecast Area Chart, 186, 192
graupel, 61
graupel-ice process. *See* cloud electrification process

235

Index

Great Blizzard of 1993. *See* Storm of the Century
greenhouse effect, 215
greenhouse gases, 215–216
Growing Degree Day, 129

H

Hadley, George, 16
hail, 62–64
    soft, 61
hailstone, 63–64
halo, 43
Heating Degree Day, 127
heavy metals, 196
Helmholtz wave, 37
high pressure area, 18
horizontal pressure gradient force, 18
humidex, 133
Hurricane
    Hazel, 160
    Opal, 160
    White. *See* Storm of the Century
hydrocarbon, 196
hydrogen sulphide, 195
hydrometeor, 57. *See also* drizzle, freezing rain, graupel, hail, ice pellet, rain, soft hail, snow, snow grain, snow pellet

IJ

ice crystal, 55, 56, 59–61
    phenomena. *See* light pillar, sundog
    theory. *See* raindrop formation process
ice pellet, 62
Ice Boat, 152
Ice Storm '98, 144–145
Intergovernmental Panel on Climate Change, 222
inversion, 89, 200
    frontal, 200
    nocturnal, 89, 200
    terrain induced, 201
isobar, 18
jets, 87. *See also* jet stream
    nocturnal, 89–90
    polar, 87
jet stream, 23, 87-89, 118–119

L

La Niña, 117–120
lapse rate, 200
    negative, 15, 200

    reversed. *See* inversion
    thermal, 14
lee cyclogenesis, 24
lift. *See also* convection
    dynamic, 27
    frontal. *See* frontal lift
    orographic, 27
    turbulent, 29
lifted condensation level, 26
light colour spectrum, 100–102
    scattering of, 101
light pillar, 106
lightning, 66–79
    See also stroke
    ball, 76
    detection network, 76, 180
    discharge process, 72
    effects of, 77–78
    hotspot, 122
    protection, 78
    stroke mechanism, 71
lightning stroke.
    cloud-to-air, 70
    cloud-to-cloud, 70
    cloud-to-ground, 70, 72
    intracloud, 70
    upward propagating, 74
looming, 108
low
    Colorado, 25, 134, 147
    Texas, 147
low pressure area, 18

M

mares' tails, 42
mesocyclone, 92, 94
mesopause, 15
mesosphere, 15
methane, 215
microburst, 91
microwave snow depth sensor, 166
Milankovitch, Milutin, 213
millibar, 12
mirage, 107
mist, 34
molecule
    catalyst, 209–210
    water, 58
Montreal Protocol on Substances that Deplete the Ozone Layer, 210

Index

N
Nipher-shielded Gauge, 165
nitric acid, 204
nitric oxide, 195
nitrogen dioxide, 195, 203
nitrogen oxide, 194, 209
nitrous oxide, 195, 215
No-name Hurricane. *See* Storm of the Century
North American Drought Monitor, 139
N-region, 69, 70, 71
nuclei
    condensation, 25–26, 55, 198
    freezing, 55
numerical weather prediction, 183–185
NWP. *See* numerical weather prediction

O
Ontario Low Water Response, 138–139
oxygen
    atomic, 209
    diatomic, 209
ozone, 202–203, 205, 208
    depletion, 210–211
    layer, 210–211
    low-level, 208
    stratospheric, 208–211
ozonesonde, 173

P
Pacific Ocean currents, 116
Palmer Drought Index, 138
parhelia. *See* sundog
particulate matter, 196, 202, 205
passive electrical phenomena, 79
Patterson, John, 170
PBL. *See* Planetary Boundary Layer
perturbation, 199
Phillips, David, 144
Planetary Boundary Layer, 151, 198
plasma gas, 80
plume. *See* pollutant cloud
pollutant
    cloud, 200
    persistent organic, 196
    primary, 194
    secondary, 194
Prairie Schooner, 152
P-region, 69, 74
p-region, 69, 71
psychrometer, 168

R
radar
    Doppler, 176
    polarimetric weather, 176
    weather, 174–176
radiation
    absorption of, 13
    incident 13
    infrared, 13
    reflection of, 13
    solar, 13
    ultraviolet, 13, 208–209
radiatively active gases, 207
radiosonde, 173
raindrop formation process, 56
    Bergeron process, 57
    collision theory, 57
    ice crystal theory, 57
rain gauge
    manual, 165
    tipping bucket, 165
rain, 57
    freezing, 62
rainbow, 102–103
    primary, 103
    secondary, 103
    tertiary, 103
red sprite, 74
relative humidity, 26
remote sensing, 174
retrogression, 22
Richardson, Lewis Fry, 162, 183
ridge, 23
rime coating, 61
riming process, 61
"ropes of Maui." *See* crepuscular rays

S
satellite
    geostationary, 177–178
    polar-orbiting, 178
    TIROS, 177
    weather, 177–179
saturated air parcel, 26
saturation vapour pressure, 54
shoreline convergence, 99
smog, 201, 203
    photochemical, 202–203
    winter, 204
snow, 58, 134
    crystal, 59
    grain, 62

Index

snow (cont'd)
  lake effect. *See* snowsquall
  pellet, 61
  pillow, 166
snowflake, 60, 146
  dendritic, 60
snowsquall, 150–151
snowstorm, 146–147
SODAR, 179
solar spectrum, 100
sonic detection and ranging. *See* SODAR
sound wave refraction, 110–111
Southern Oscillation, 118
St. Elmo's fire, 80
standardized observing protocols, 163
steering flow, 22
stepped leader, 71
Stevenson screen, 164
Storm of the Century, 147
storms. *See also* Alberta Clipper, blizzard, Ice boat, Prairie Schooner, snowsquall, snowstorm, thunderstorm
  autumn, 160
  ice, 144–145
  synoptic scale, 22, 59
  summer, 142,
  winter, 143, 144–152
stratopause, 15
stratosphere, 15
streamer, 71
subsidence, 27
subsiding air, 23
suction vortices, 94
sulphur dioxide, 195, 204
sundog, 105. *See also* diamond dust
"sun drawing water." *See* crepuscular rays
sunshine hours, 141
supercell, 92–93
synoptic meteorology, 19–20

T

TAF. *See* forecast, Terminal Aerodrome
temperature scale
  Celsius, 166
  Fahrenheit, 166
terminal velocity, 56, 57
thermals, 90
thermistor, 167
thermometer, 166–167
  dry bulb, 168
  maximum, 167
  minimum, 167
  wet bulb, 168

thermosphere, 15
thunder, 72, 74, 111
  pitch, 74
thunderstorm, 152–155
tornado, 53, 92–95, 156–159
Tower of the Winds, 170
trade winds. *See* winds
tripole distribution, 69
tropopause
troposphere, 15
trough, 23
trowal, 28
turbulence, 89, 199
turbulent flow, 89
turbulent mixing, 16

UV

updraft, 93
urban heat island, 127
vapour pressure, 54
virga, 64
volatile organic compounds, 196, 202, 205

W

warning program, 186
waterspout, 95
water vapour, 215
wave
  long, 21–22
  planetary, 22
  short, 21, 22–23
weather
  forecasting process, 182
  radar, 174–176
  satellites, 177–179
  space, 80
  warning, 186
  watch, 186
wind chill, 132
wind flow, 199
wind patterns, effects on. *See* Coriolis effect, curvature in air flow, frictional effects
winds, *See also* jets
  anabatic, 86
  convection-induced, 90
  katabatic, 86
  onshore, 99
  plow. *See* downburst
  solar, 80
  terrain effect. *See* anabatic, katabatic
  trade, 16, 116–117
  vane, 170

Photo Credits

Arnold Ashton 190; Canadian Wind Energy Association (a Canadian wind farm) 83; Keith Chadwick 240a; Phil Chadwick 1, 5, 22, 25, 30, 31, 32, 33, 34a, 34b, 35a, 35b, 36, 37, 38, 39, 40, 42a, 42b, 43, 44, 46, 47, 48, 49, 51, 95a, 97, 99, 103, 104a, 106, 108, 112, 129, 142, 143, 145a, 146, 148, 150, 153, 166, 169b, 171b, 181, 183, 192, 201, 230; The COMET Program 12, 57, 87; Chris Cowan 163; Creative Commons, Attribution Share Alike: Luna04 78b; Creative Commons, Attribution Share Alike: United States Air Force / Senior Airman Joshua Strang 81; Environment Canada 50, 84, 118, 119, 125, 126, 128, 130, 133b, 135, 136, 137, 138, 141, 145b, 149, 157, 158a, 158b, 164, 165a, 165b, 167, 168, 169a, 172, 173, 175b, 178a, 178b, 179, 185, 194, 195, 203, 204, 205, 217, 218a, 218b, 219a, 219b, 220a, 220b; Environment Canada / Steven Brady 175a; Environment Canada / Geoff Coulson 159; Environment Canada / Dave Sills 52, 154a, 161; Environment Canada / unknown photographer 90; Ron Gravelle 61b; Bill Hume 240b; iStockphoto.com / Hans Caluwaerts 210; iStockPhoto.com / Slavoljub Pantelic 213; iStockphoto.com / Shane White 211; iStockphoto.com / Serdar Yagci 187; JupiterImages Corporation 10, 100, 156, 193, 206, 224; Randy Kennedy 189; Alister Ling 45, 72, 78a, 82; Edward P. Lozowski 64; Meteorological Service of Canada - Edmonton / Bill Burrows 77, 122; NASA 207; NationalAtlas.gov (adapted by Lone Pine) 7; NOAA Photo Library 61a, 66, 147; NOAA Photo Library / Grant W. Goodge 41a; Dave Patrick 155b, 155c; PhotoDisc, Inc. 75; Rob Pigott 96; public domain 76, 170, 171a; The Weather Network / David van Gurp 92; Gary Whyte 54

Authors

Phil Chadwick

Trained at Queen's University as a nuclear physicist, "Phil the Forecaster" has been a professional meteorologist since 1976. Phil worked from Edmonton to Halifax with several stops in between. Most of his meteorological career has been in Ontario as a severe weather meteorologist and researcher. Phil was an instructor with Training Branch of the Meteorological Service of Canada in the mid-1980s where he first became known as "Phil the Forecaster."

Phil has always been an artist and passionate environmentalist. When he is not at work, Phil paints "en plein air" and maintains an apiary. His art and honey have found their way around the world. Life is good—plus you gotta laugh!

Bill Hume

Bill started his career in atmospheric sciences in 1971 with Environment Canada in Toronto. He attained a Master's degree in meteorology from the University of Alberta in 1975 and has spent most of his working career in Edmonton and Calgary. He has experience in most aspects of meteorology including weather forecasting, climatology, air quality science and atmospheric research. He has taught meteorology at the University of Alberta and served on several national committees including the Canadian Meteorological and Oceanographic Society. Bill and his wife Judy have two married children and live in Edmonton.